ADVANCE PRAISE FOR **Privacy** AND **Philosophy**

"Contemporary privacy issues tend to be discussed in legal, policy or sociological terms. McStay adds a welcome philosophical context to this discussion. Impressively erudite, *Privacy and Philosophy* takes the reader on a trans-century tour that enlarges our understanding of the idea and its implications."
—*Joseph Turow, The Annenberg School for Communication*

"More than at any other time in recent history we are confronted with the pressing questions and contradictions raised by the notion of privacy—and McStay's brilliantly illuminating philosophical tour of the concept provides thoughtful and original answers that will serve as touchstones for discussions of privacy in the era of Facebook, NSA data mining and beyond."
—*Mark Andrejevic, The University of Queensland*

"The book gives a very original and kaleidoscopic perspective on the notion of privacy in an age of social and ubiquitous media. The well-chosen selection and in-depth discussion of evident and less evident philosophical views broadens and deepens the view on this timely and intensely discussed issue. Especially the framing of privacy as an affective set of protocols within the social realm offers relevant and refreshing insights."
—*Jo Pierson, Associate Professor, Vrije Universiteit Brussel (iMinds-SMIT)*

"Offering a fresh and authoritative take on an established concept, McStay avoids the trap of only asking what philosophy can tell us about privacy, but also considers what privacy can tell us about epistemology, ontology and metaphysics. This is an important contribution to our understanding of how privacy and publicity operate in culture today."
—*Clare Birchall, King's College, London*

Privacy AND Philosophy

Steve Jones
General Editor

Vol. 86

The Digital Formations series is part of the Peter Lang Media and Communication list.
Every volume is peer reviewed and meets
the highest quality standards for content and production.

PETER LANG
New York • Washington, D.C./Baltimore • Bern
Frankfurt • Berlin • Brussels • Vienna • Oxford

Andrew McStay

Privacy AND Philosophy

New Media AND Affective Protocol

PETER LANG
New York • Washington, D.C./Baltimore • Bern
Frankfurt • Berlin • Brussels • Vienna • Oxford

Library of Congress Cataloging-in-Publication Data

McStay, Andrew.
Privacy and philosophy: new media and affective protocol / Andrew McStay.
 pages cm. — (Digital formations; vol. 86)
Includes bibliographical references and index.
1. Privacy, Right of. 2. Privacy. 3. Digital media—Social aspects.
4. Information technology—Social aspects. I. Title.
JC596.M36   323.44'801—dc23   2014005855
ISBN 978-1-4331-1899-9 (hardcover)
ISBN 978-1-4331-1898-2 (paperback)
ISBN 978-1-4539-1336-9 (e-book)

Bibliographic information published by **Die Deutsche Nationalbibliothek**.
Die Deutsche Nationalbibliothek lists this publication in the "Deutsche
Nationalbibliografie"; detailed bibliographic data is available
on the Internet at http://dnb.d-nb.de/.

The paper in this book meets the guidelines for permanence and durability
of the Committee on Production Guidelines for Book Longevity
of the Council of Library Resources.

© 2014 Peter Lang Publishing, Inc., New York
29 Broadway, 18th floor, New York, NY 10006
www.peterlang.com

All rights reserved.
Reprint or reproduction, even partially, in all forms such as microfilm,
xerography, microfiche, microcard, and offset strictly prohibited.

Printed in the United States of America

Contents

Chapter One: Introduction   1

Section One: Living Together   13
Chapter Two: Aristotle, borders and the coming of the social   15
Chapter Three: Liberalism, consent and the problem of seclusion   22
Chapter Four: Utilitarianism, radical transparency and moral truffles   36
Chapter Five: Pragmatism: Jettisoning normativity   49

Section Two: Knowing   61
Chapter Six: Heidegger (Part 1): Concerning a-historical being and events   63
Chapter Seven: Heidegger (Part 2): On moods and empathic media   74
Chapter Eight: Latour: Raising the profile of immaterial actants   88
Chapter Nine: Phenomenology: The rise of intentional machines   102
Chapter Ten: The subject: Caring for what is public   107
Chapter Eleven: Alienation: The value in being public   123
Chapter Twelve: Spinoza: Politics of affect   134
Chapter Thirteen: Whitehead: Privacy events   143
Chapter Fourteen: Community facts   151

Appendix: An A to Z of privacy: New theories and terminology   163
Notes   167
References   171
Index   181

CHAPTER ONE

Introduction

A life without privacy is impossible. It connects with the most basic processes of how we live together and the institutions we create, and we can be absolutely clear upfront—there is no question of privacy disappearing. The notion that privacy might somehow be removed, surpassed or lost from the human equation is to very much misunderstand it. This is because privacy plays a fundamental role in our most basic daily interactions. Be this in our behavior towards each other, what we consider to be taboo, our modes of intimacy, the confidences we share with others, how we arrange our homes and working spaces, where we store thoughts and things of value, and more recently the ways that these are imbricated in media and technological systems, privacy is a very basic and primal premise. While passionately argued and defended, privacy is one of those words that does not lend well to precise definitions. In general though we use it as a means of referring to borders (keep out!); as a means of maintaining dignity in human behavior (for example sex or defecation); to highlight autonomy and the right to control aspects of our lives, relationships and bodies; and as a way of addressing the security of information that concerns us in some way. Like words such as freedom, liberty, equality and rights, we feel privacy to be an important ideal although laypeople and professional academics struggle to delineate 'it.' Privacy International (2013), a charity whose stated aim is to defend the right to privacy across the world, defines privacy as:

> ... the right to control who knows what about you, and under what conditions. The right to share different things with your family, your friends and your colleagues. The

right to know that your personal emails, medical records and bank details are safe and secure. Privacy is essential to human dignity and autonomy in all societies. The right to privacy is a qualified fundamental human right—meaning that if someone wants to take it away from you, they need to have a damn good reason for doing so.

There is no univocal meaning but privacy is better thought of as a rallying-word for a cluster of interests. In the quote there are a number of key words that have philosophical resonance, these being: control, dignity, right and autonomy—all highly liberal notions. Privacy International is accurate in that privacy is a qualified right. This means it is not (in liberal terms) absolute like other rights (for example, not to be tortured), but rather it may be encroached upon if there are other competing interests. Another key point about privacy is that while we might assume that we have less opportunity to be private than in the past given the rise of data-based technologies, this is not necessarily true. Despite technical innovation in media, feedback, profiling, targeting, bureaucracy, prediction and surveillance technologies, many societal changes since the industrial revolution involve a net increase in privacy. Be this less familiarity with our neighbors, more geographically dispersed family arrangements, working away from home, weakening of religious authority (and all-knowing deities and confessional practices), greater possibility of children having their own bedrooms, increase in car ownership (versus public transport) and so on, in many ways we are more private then ever (Westin, 1984 [1967]). Early on, we should also query the notion that privacy is essentially a positive notion. Areas of feminist discourse on privacy for example complicate over-simplified ideas about privacy as an always-desirable outcome, not least because of the potentially repressive dimension of privacy that may act as a control mechanism maintaining imbalanced power relationships (Rössler, 2005; Allen, 2011). Indeed, historically 'the family home was a man's castle but a woman's place' (Allen, 1988: 63) and patterns of female home ownership and inheritance rights remain patchy. If denied control over the home, privacy may also be a negative condition that acts as a hermetic seal against public visibility so to facilitate unaccountable behavior. On the domestic sphere, the Marxist feminist MacKinnon (1989) argues that privacy represents a domain where women are both deprived of power and recourse to legal protection, due to the historical unwillingness of the liberal state to intervene in private domains.[1] For MacKinnon this is why feminism has to make the personal political, blur private/public distinctions and make sinister Warren and Brandeis' (1984 [1890]) maxim, the 'right to be left alone.' This raises difficult questions about the reach of the state, with many liberal feminists recognizing historical imbalances but maintaining that privacy remains fundamental, particularly given its association with autonomy that facilitates richer, deeper and more significant relationships away from the attention of others (Allen, 1988; DeCew, 1997).

Journalism also complicates the privacy-as-positive narrative by means of its overriding interest in making information public, transparent and open. Like professional fighters in the ring, on the one side we have 'the right to be left alone' and the other, the 'public's right to know,' and the latitude given to journalists to conduct public interest investigations that may involve intrusion (Lloyd, 2012). On privacy invasion in journalism, Alan Rusbridger (2012: 142), editor of *The Guardian*, suggests five conditions: there must be sufficient cause along with prior assessment of harm to individuals and families; integrity of motive and justification that public good will follow; methods should be proportionate to the story and degree of public interest, and intrusion minimized; intrusion should be overseen by an authority; and there should be reasonable prospect of success with 'fishing expeditions' not justified. However, privacy in journalism by means of secrecy *is* necessary in the case of whistleblowing and the protection of journalists' sources, and somewhat paradoxically we cannot promote transparency without privacy.

Propositions to be assessed

The point of this preamble is to highlight that privacy is deeply implicated in our lives at fundamental levels, that it is not solely positive or neutral, and that it has been broached in many ways, particularly in relation to new media technologies. Less however has been said about the philosophical dimensions of privacy. This is the subject of this book. While lacking the exact focus of empirical work or dedication to a sole topic such as 'mobile apps, teenagers and privacy perception,' my intention is to bring about new ways of thinking about privacy by casting a wider net to account for privacy. The vast literature on privacy in journal and book form affords some risk-taking, experimentation and unique ways of coming at privacy—which I employ particularly in the mid-to-latter half of this book. As so much has been written about the implications of novel technologies and their effects on social arrangements, I am reticent to recap all of this (for overviews see Lyon, 2001; Petronio, 2002; Rössler, 2005; Bennett and Raab, 2006; McStay, 2011). Instead this rich context of study provides latitude for something different. While early chapters assess familiar philosophical perspectives that underpin the majority of assumptions of privacy studies, later chapters are more novel in orientation. In a sense, this is to play at dressing-up, to try different hats and outfits, to mix and match at times, and see how privacy looks if we depart from the usual dualistic and liberal approaches that inform privacy studies. There is both a frivolous and more earnest element to this, but both aim to refresh our understanding of privacy, and provide new fields of inquiry and ways of thinking about privacy. While this book draws on a diverse range of approaches to explore the relationships between privacy and philosophy, particularly as they apply to media, I will develop my own arguments in dialogue with these philosophical traditions, particularly in reference to the following propositions that will be unpacked and referred to in this book. These are:

1. Privacy should be conceived in terms of affective events;
2. Privacy is an emergent protocol that contributes to the governance of interaction among people and objects.

Ideas about affect have a long philosophical history encompassing a wide range of thinkers and tactics that will be addressed in the course of this book. In general however an affective approach prefers to understand life in lived, immediate, experiential, sensational and felt terms. Important too is the recognition that an affective event is that which has the capacity to transform and influence other actors around it. This may involve inter-personal dynamics in terms of how two people relate and possibly cohabit, or global communication structures and international law. While today privacy is often discussed in terms of Facebook, Google, media behemoths and governmental surveillance, these are only small footnotes in a much larger story about how privacy comes to be by means of events that are to be recognized as felt and lived, as well as theorized.

Clearly an affective approach cannot be all-encompassing because privacy is regularly breached without us being aware. My second proposition on protocol and governance is a response to this and asserts that privacy can be usefully conceived in emergent and ecological terms. I am mindful that the latter is a somewhat trendy expression, but as the science of interconnection it characterizes well my view of privacy that has less to do with being alone, but that privacy norms contribute significantly to how we connect and interact with others. A systemic approach thus sees privacy as an organizational principle that contributes to the regulation of institutions, practices, modes of interaction, and social and individual life more generally. As a principle, or set of principles, privacy is best conceived in terms of protocol informed by physical, social, historical, technological and environmental circumstances. Importantly, protocols are not imposed, but are co-created between us and emerge out of arrangements comprised of dynamically interacting parts, the creation of novel groupings and arrangements, the inclusion of new actors, and emergent norms that govern interaction and behavior. I am not first to arrive at this type of idea as Petronio (1991, 2002) has written extensively on privacy and management, and Altman (1975) offers ecological observations in regard to the coherence of privacy as a system that regulates behavior between people. Benn and Gaus (1983) also make claims on privacy as that which regulates social behavior, and that as norms and circumstances change so will patterns of privacy behavior.

Later and more uniquely however I will develop this to account for the role of technical objects and media systems, and the means by which they are connected to people to form a more expanded view of social groups and emergent norms.

A basic argument of this book is that a systemic view of privacy excludes the possibility of conceiving of privacy as being alone and if we are to think in terms of seclusion, this must be in terms of modulating relationships with others. Indeed, the fact that privacy involves the modulation of connection with others is only not listed as a key proposition because it is subsumed within the second proposition on protocol. Being alone and being private are different as the former involves an absence of relationships and connection. This absence can be quite palpable and may even occur in public as well as on a hypothetical desert island where no one knows nor cares that an unlucky individual is stranded there. Without connection, being alone is simply that—absence of others that relate to us in some way. Privacy works differently as it involves entities (including people, animals, systems and organizations) actively managing relationships with each other. This involves the establishing of systemic norms, akin to a script, or that 'mediating device that draws together theatrical, psychological, and machinic roles' (Fuller and Goffey, 2012: 158). Applying the notion of a script to *privacy protocol* we can see the way in which privacy scripts frame the terms and parameters by which elements of a system interact and behave. While I do use the word 'norms' in this book, I prefer the term 'protocol' as it has less to do with framing privacy in terms of moral obligation, commandments and recommendations. As will be developed, I see privacy in more basic, non-rational, behavioral and affective terms. Protocol also more readily involves prescriptions that inform and guide relationships between humans, non-humans and systems that people contribute to and interact with.

By discussing privacy in terms of scripts, protocols and indeed norms, this indicates that we should be able to discern public objective features. This is a fair but difficult point. If we were to discuss happiness we would not be able to describe these without recourse to behavioral traits, neuroscientific firings and interrelationships between people. As to what happiness itself is, this is an unanswerable question, although we do not deny its existence. The same goes for privacy in that it is a protocol that depends on its actors for existence. At the systemic level we can open this out further to again see privacy in ecological and connective terms, and the protocol (scripts) that exist to regulate the activities of organizations or arrangements. The invocation of ecology is not just a metaphor, but sincerely employed as privacy involves self-creating systems of connections, inter-actions and happenings to emerge from these connections. As existing communities develop greater and more intense means of connection we see novel social organisms and behavior emerge, and new means of community governance and regulation.

A memo on method

Why philosophy? Indeed, why take recourse to an examination of privacy by means of philosophy, particularly when some of the philosophers broached in this book do not even deal with privacy. The answer is that it helps us deal with the terms by which we ask questions, conduct empirical research, create policy, fashion and enforce regulation, and assists with understanding the construction and reasons for our norms and principles. In general it helps challenge assumptions and while philosophy may confuse and obscure what was once clear, it also problematizes wrong-headed assertions and forces us to re-evaluate our position in regard to fundamental issues. So while this book is not of an empirical nature, I am very interested in method and guiding principles of enquiry. This book, then, is an examination of key perspectives and orientations that do, or may, inform the study and conceptualization of privacy. One hope for this book is that it may act as a reference point for researchers to check the roots of their own understanding of privacy and question their orienting principles. If the roots of a particular approach are not covered in this book (and I make no claim to be comprehensive), privacy researchers might similarly uncover and assess the steering philosophical ideas that inform their own operationalization of privacy before heading into the messy world of people, politics, law, security, information management, computer science, surveillance, business, media and other domains involving privacy matters. This is not to suggest however that philosophy is a set of master narratives somehow more important than other areas of study and research. This most definitely is not the case, but philosophy is central to understanding the historical context of ideas, and performs as an irritant to ideas occupying status quo. Philosophy has an annoying habit of forcing us to explain the terms of our assumptions and theoretical propositions, yet at its most joyful offers systematic demolition of bullish arguments and false rhetoric.

The choice of philosophers, schools and approaches in this book reflects both dominant discourses within critical and/or empirical literature on privacy, opportunities for fruitful interpretation of privacy, or a deep involvement in matters associated with privacy. The line-up to be broached includes: Greek dualistic accounts and liberalism because in relation to privacy these pass for common sense; utilitarianism and pragmatism because of their stated rejection of deontology; Heidegger, Latour and phenomenology both because of connections with technology, but also because of their influence on anti-substantialism and anti-foundationalist critique; Ryle and Wittgenstein because they make public what was thought to be the preserve of the private; Hegel and Marx because (by means of Kant) they underpin critical theory; Spinoza and Deleuze for their recognition of affect; and finally

Whitehead due to his far-reaching account of context and events, and because a Whiteheadian ontology informs my overall account of affect and emergent protocol. While clearly a wide-ranging line up, each provides a rich supportive seam with which to understand approaches to privacy.[2]

Structure of the book

The book is split along the lines of ethics (how we should live) and epistemology (concerned with questions about what knowledge is, where it comes from and what it is to know something). Early chapters (2–5) in Section 1 focus on ethical dimensions and the ways in which privacy norms contribute to *how we live together*. These are political in orientation having to do with questions about the relevance of classical public/private distinctions in a media saturated age, questions about the construction of freedom and autonomy, and an assessment of utilitarianism and pragmatism. Chapter 2 makes sense of what I designate a border-based conception of privacy that is Aristotelian and Greek in origin from around the time of the Peloponnesian War (431–404 BC). This involves the first clear formulation of privacy norms in terms of public civic life and private life, and also as a negative phenomenon with all positive value being reserved for public life. While the negative view of privacy is less prevalent today, this dualistic conception maintains its hold as we conceive of privacy in binary terms. This chapter progresses to discuss the implications of Arendt's (1998 [1958]) critique of Greek public/private distinctions by means of the rise of the social. It also highlights Arendt's glaring omission of media in her account of public/private implosion and concludes by making those links on her behalf.

Chapter 3 explores liberalism arguing that both critics and defenders of privacy may find the roots of their arguments in liberalism. This chapter thus offers origins, criticisms and an assessment of ways in which its principles feed into contemporary thinking about privacy. Mindful of aforementioned cases where privacy facilitates wrongdoing, it broadly agrees with the liberal sentiment that privacy is positive and that in contrast to parrot-like statements about 'having something to hide,' privacy is better regarded in terms of control, dignity and respect for self-hood regarding what one wishes to share, reveal or allow access to. Notably, this involves being open as well as secluded. While a high proportion of accounts of privacy in the wider literature employ a broadly liberal approach, a thoroughgoing account is needed because too often there is amnesia and uncritical acceptance of its root suppositions. This chapter also takes some of those assumptions to task.

Chapter 4 explores utilitarianism by means of Bentham, Mill and Richard Posner, whose divisive account of privacy is deeply influenced by Bentham's teleological writing and enlightened interest in transparency. Indeed, this focus on teleology (or consequentialism where ends justify means) reveals a deep split between Bentham and Mill, particularly evident in Mill's *On Liberty* that continues to indirectly act as a manifesto for privacy rights and watchfulness of governmental involvement in our lives. This chapter develops so to explore and critique Posner's Benthamite account predicated on *radical transparency* of its citizenry. This involves an inversion of enlightenment ideals, which sets the scene for Heidegger to be discussed later.

On pragmatism, Chapter 5 assesses ethical stances that pertain to looking forward rather than backward. This is that creative area of philosophy that revels in indeterminism and refuses to be restricted by prescriptive normative dogma, strictures and teachings from the past. Indirectly pragmatism feeds systemic and contextual accounts. This leaves norms to be decided by actors themselves in relation to the specifics of the situations they find themselves in. Paying particular attention to Richard Rorty and key influences of his own (John Dewey and William James), this chapter foregrounds Rorty's anti-foundationalist critique and its connection with Nissenbaum's (2004, 2010) contextual view of normativity and privacy. While both pragmatic and contextual approaches have attraction and are influences on my own thoughts about protocol, questions remain over what is lost by a rejection of deontic values and whether past ethical foundations should be jettisoned so readily. Might a pragmatic and contextual view of privacy be dangerous?

Section 2 on *Knowing* (Chapters 6–13) departs from ethics, politics, social arrangements and how we should live together, to examine what is known by systems said to be privacy-invasive along with consideration of the increasingly closer relationships between people, machines and mediated experience. If privacy is intimately linked with knowledge, what is its connection with epistemology? Further, what might contemporary informational privacy matters teach epistemology about the contemporary state of knowledge? This section also investigates more thoroughly the ontological status of privacy and the ways in which privacy itself can be said to be an actor by means of its capacity for affect.

Chapters 6 and 7 are both dedicated to examining Heidegger, technology and technical rationality in terms of novel developments in the field of behavioral advertising. Heidegger's argument suggests that technical answers to privacy and technological problems will not get us anywhere, but rather we need philosophical understanding to combat what he phrases as a metaphysics of presence, or those preeminent ontological discourses about quantity, efficiency, utility, reserves, substances, exchange, accumulation, stockpiling, being productive, producing more cheaply, getting faster, going further, and so on. As mentioned above in relation to Bentham and transparency, Heidegger is valuable because he helps us understand the will to make all aspects of life visible and usable. These chapters also introduce Heidegger's notion of the 'event,' relying on both my own primary reading and Harman's (2002, 2007) novel interpretation of Heidegger that clarifies what are difficult, unfinished, highly obscure, yet valuable ideas. I have run a risk in including Heidegger's discussion of events because while his approach connects to the notion of events in other philosophers invoked (most notably Whitehead, in Chapter 13), ontologically-speaking it is not a perfect match. To be clear, the premise of affective events stated in proposition 1 is closer to Whitehead. The key point I seek to relay about Heidegger's events in this chapter is that the event is an *opening for disclosure* of how things come to be and how they are for us. My argument in relation to Heidegger is that technology increasingly gives the appearance of understanding the distinctiveness of our lived situations, and the 'thisness' or 'suchness' of situations as we encounter them in a mediated setting. This sees technology as not only technical, but as engaging with us at non-technical levels. I argue that the possibility that technical systems might engage with us in such a seemingly intimate manner is characterized by *empathy*, or that ongoing ability to pick up and act on the emotional state of others, what is significant for others, along with their intentions and expressions. Such connections with data mining systems involve close and longitudinal assessment of our dispositions and behavior, and what in this chapter is formulated as co-creatively authored moods of information, or co-emergent 'being-in-the-world.' This is not to argue that technical systems have access to some interior private 'us' (which itself is problematized in this book), but rather that systems are increasingly capable of verisimilitude and able to pass-off a sense of knowing and understanding by means of their predictive capacities. Verisimilitude in relation to behavioral systems is a point I develop in some depth in this chapter and have done elsewhere (in *The Mood of Information*, 2011). It involves appropriation of Bateson's (2000 [1972]) remark that the difference between being right and not wrong is not distinct. I argue that our relationships with technical systems are best understood in terms of *co-evolving authorship* and those modes of autopoietic feedback relationships that create more of the same assemblage (for example, more highly refined targeted advertising begets more of the same by means of co-development between people to be exposed to advertising, their online behavior, machines, advertising networks, data processes, advertising agencies and so on). For those interested in privacy and/or advertising, it is important we place critical emphasis on assessment of the means by which collaborative and co-evolving authorship between people and systems occurs, and more thoroughly understand the nature of the feedback relations therein and implications for privacy conceived both in systemic and affective terms.

Chapter 8 on Latour also examines our relationships with technical systems although the purpose and tenor of this chapter is quite different to that which precedes it. Here I argue that privacy may be considered as an 'actant' in its own right and can be defined as an outcome with the capacity to affect that around it. This is an ontological assertion about the status of privacy, but for this to be a reasonable idea some groundwork is required. This is done by means of Latour who argues for a coming together of social, economic, political, technical, natural and scientific entities on the same terrain as each other so to form a flat ontology (Harman, 2008). Reality in this line of argument is not dictated by materiality, substance or metaphysics, but the capacity to affect. Substance in Latour's view is to be jettisoned in favor of process, assemblages and what he designates 'black boxes.' His interest is in the stability of things, but these need not be corporeal. This allows social phenomena the same ontological rights as things or elements from the Periodic Table. It sees a downplaying of genus, kind, sorts and types in favor of assessment of the processes that take place across these scales, and how they interact and affect each other. In a philosophical place characterized by hybrids, transduction and assemblages of people, things and processes, we can now begin to see privacy as an outcome and co-articulation of interacting entities from a variety of scales. By removing the need for substantialism to properly exist in favor of the capacity to influence and affect, privacy protocol comes into more discernable relief. As both protocol created by entities from interacting scales and as an affective outcome when protocol is breached, privacy is able to both regulate and transform that around it. More practically, this capacity for affect is defined in terms of the capacity to bring about change in behavior whether this be at the level of the individual, in the home, in the workplace, in hardware and software systems design, in policy and legislation, in product design, in corporate strategy or in the design of national security policy. This leads to the argument that because of its capacity for affect, privacy is both an assemblage and actant in its own right—or that which is capable of modifying the state or behavior of that around it.

Having introduced Heidegger earlier, I have already done some strange things with phenomenology and Chapter 9 on phenomenology itself continues in this vein. Specifically it offers a systems-based reading of Brentano and Husserl in regard to their notion of 'intentionality,' or that key phenomenological premise which refers to the structuring of experience and how things appear so to provide an experiential unity. In general intentionality involves being directed towards something and refers to human experiences of objects. What, then, if we ask about the possibility of *machinic intentionality*? In the chapter on Heidegger I posit the idea that machines may have empathic qualities. If we can accept this, then what are the objectifying processes for machines? We cannot conclusively answer this, but by means of asking these types of questions we can more clearly recognize that machines involved in the soliciting and handling of information are not simply props to our lives, but are discerning entities in their own right.

Chapter 10 addresses both the nature of subjectivity in relation to technical assemblages (made of subjects, objects and processes) by means of Wittgenstein, Ryle and Russell. It pays attention to the proposition that what we consider as the liberal sovereign subject with its interior self bordered-off from the world is not as secure as first thought. Could it be that much of what we consider as constitutive dimensions of self and character are actually more public than private? This is to ask a basic question about what it is to know others and what this knowledge is constituted of. This chapter adds an additional question asking whether objects as well as subjects are capable of knowing, if the capacity to know others exists at all? These are important discussions because if what we believe to be interior is more exterior than first thought, what is public takes on renewed significance, quite possibly requiring a re-evaluation and re-estimation of the value of that which is public.

Chapter 11 continues with an interest in the self, but this time examines relationships between subjectivity, critical theory, Hegel, Marxism, alienation and privacy. This is a difficult relationship and while much work on critical political economy, media and labor invokes neo-Marxist notions such as Dallas Smythe's audience-as-commodity thesis (Smythe, 1981; Bermejo, 2009; McStay, 2011; Fuchs, 2012), they do not account for the full implications of a Marxist-inflected account of privacy. Both theoretically and historically this is deeply uncomfortable, as Marx directly attacks the premise of human rights. Privacy is also problematic because of the connection with property. After all, what need is there of privacy in truly communal life? However, an examination of the Hegelian and Marxian notion of alienation provides a highly usable critical tool without having to employ all aspects of Marxian theorizing. Indeed, in a media context, attention might be more productively focused on alienation itself than labor.

Chapter 12 addresses a point sorely missing in systems and protocol-based accounts of privacy by means of the notion of affect. This is done, possibly unusually, without the normal route through Foucault (1990 [1976]) and his ideas about biopower and biopolitics, but by means of a few historical steps backward to examine the influence of Spinoza. The reason for this is that I seek to connect earlier discussion about a Latourian flat ontology with the parallel theorizing of affect that refuses the idea that mind and body are totally distinct. In developing a sense of the tangible dimensions of privacy I define these in terms of *phenomenal materialism*—or that lived sense of the material world in which we reintroduce the 'being' part of being private back into understanding.

Dwelling on affect, Chapter 13 assesses the influence of Whitehead for privacy discussion. There are two parts to this: first is to discuss Whitehead's own account of privacy that has to do with the formative stages of processes before they become public; the second (and more central point to this book) is to account for Whitehead's cosmology of events and process as a means of understanding the raw underpinning logic of dynamic approaches to privacy, while also being able to fold affect into the account. Whitehead is a 'big picture' philosopher, but his worldview is also practical and encompassing, particularly given the overall presentation of privacy in this book as that which comes to be by means of processes, interactions, affects, and the creation of new arrangements and ways of doing things.

In concluding the book, Chapter 14 briefly recaps and highlights the most appropriate ways in which to conceive privacy. It also concisely details key outcomes, approaches, recommendations and arguments generated. Lastly, key analytical terms and novel arguments have been italicized in-text and briefly defined in the appendix. This A to Z of approaches to privacy (admittedly with a few letters missing) may also serve as a shortcut to the book for those wanting to get to the point quickly without taking in the scenery.

SECTION ONE
LIVING TOGETHER

CHAPTER TWO

Aristotle, borders and the coming of the social

Border-based understanding dominates our phrasing and conceptions of privacy matters. It involves a public/private duality and can be connected with other pairings that inform the ways in which we formulate an understanding of privacy. Be this inside/outside, internal/exterior or intimate/distant, we have pairings that are seemingly fundamentally different in nature, if not oppositional. In lived life however pairings need a zone, object or means by which to tell the two domains apart. While borders may be fixed as with walls, fences and sunglasses, they may be less rigid instead involving behavioral boundaries such as averting one's gaze, managing proximity to others, maneuvering without touching others, or the use of decorum and emotional restraint as a means of initiating virtual walls. A border-based conception may thus involve material and architectural norms, as well as social constructions of a behavioral and less substantial nature. While the existence of privacy is universal across people, its make-up differs enormously, and borders are deeply ethnocentric and dictated by agreed cultural norms (Ford and Frank, 1951; Hall, 1969; Moore, 1984; Westin, 1984 [1967]; Mead, 2001 [1928]; Malinowski and Ellis, 2005 [1929]). This chapter explores in greater depth the roots of border-based privacy in relation to Aristotle and Greece from the time of the Peloponnesian War (431–404 BC). Following Moore (1984), the specific period and place in question is Athens from the end of the Peloponnesian War (404 BC) to its defeat by Philip of Macedon in the battle of Chaeronea (338 BC).

It is in this period that we see the beginning of the recording of the Greek language, Homer, and recognizable origins of the distinctions between public and private. This dualism has spilt into both popular privacy lexicon and more recent philosophy, most notable in Arendt who depicted the implosion of public/private distinctions leaving only an expanded sense of the social that, as a result, deeply complicates these pervasive dualisms. The second half of this short chapter assesses Arendt's arguments on privacy querying the curious absence of media in her account. This is a point I address only very briefly both in terms of mass media and marginally more robustly in relation to recent media developments. Accounts of mass and contemporary media are abundant enough, so I will restrict myself to solely problematizing the border-based language of privacy by accounting for media in terms of enlargement, intensification of visibility and also the possibility of noopolitical theorizing of media.

Living the good life

Ancient Greek conceptions place great emphasis on virtue in citizenship and being public. What is good is not found in being private, but rather in being public and part of a community. Moreover, the state is not separate from the individual, but rather completes the individual. In ancient Greece, one could only properly enter public life if one was unrestrained by domestic life, owned one's property and had private affairs in order. Control over private domains and access to the public involved mastery of necessity and ruling over others so to move freely in public and political realms. While the political realm saw dialogue among equals, the private sphere of the household saw gross inequality. Freedom in this border-based arrangement equated to not being tied to private life and being free to transcend it. The 'good life' or political life was positive insofar that one had mastered the necessities (biology and reproduction, labor, finance and that which is tied to survival), so to pursue more noble ends. Likewise, being public involved notions of logic, debate, the arts and culture, and what is private was identifiable in terms of material and biological needs and wants. This split was clear: none were permitted to enter the political sphere who were tied to activities involving making a living or sustaining life processes of others. The private then was connected to animality and thereby held in disdain.

Polis and oikos

To be private stems from *idios*, or in noun form *idiotes*, which means to be occupying a private role rather than one in public office (Moore, 1984). With connection to 'idiot,' it also means a layperson, someone lacking professional knowledge or an individual considered ignorant. In general, to be private is to lack participation in civic life. The carving up of the public and the political (*polis*), and the private and domestic (*oikos*), is well evidenced in Aristotle's *Nicomachean Ethics* (2009 [350 BC]) and *Politics* (1995 [350 BC]). As Aristotle depicts in *Politics*, the highest domain is the *polis* and this ancient institution is a key forerunner of today's public sphere. In ancient Greece it was the province of political activity—with free, but very visible men, being the only constituents. Indeed, happiness for Aristotle is defined in large part by means of civic status and being able to engage in activities that allow one to flourish, thus leaving a question for Aristotelian scholars of whether a woman's life is to automatically have happiness disqualified. Citizenship for Aristotle is defined both in terms of participation and also having achieved self-sufficient existence. The *oikos* represents the private sphere, the household and family life that defined the role of women, children and slaves, as well as the masterful nature of the male citizen.

One function of the *polis* was to protect against the futility of lives lived privately and individually (and thereby characterized by loss). What is valuable is what we have in common with those before, now and to come, with the consequence being that the *polis* or public sphere takes on characteristics of transcendence made manifest on earth. Privacy by contrast refers to dimensions of life that do not fit this a-temporal vision. The Greeks had a point as after all, 2000 years later, many of their ideas discussed in their public *fora* continue to directly and indirectly shape politics and philosophy. The *polis* or city is that which exceeds its parts and while discussed by Aristotle (1995 [350 BC]) as a compound entity, the city surpasses its constituent elements. The city comes before either the family and the individual (and therefore the home or private realm) because the 'whole is necessarily prior to the part' (ibid: 11 §1253a18) and in the same way that if a body were to be destroyed, there would be no hand or foot worth speaking of. The city precedes individuals and any possible interest they might have in seclusion, because to be isolated is to be unable to engage in public political association and consequently they 'must therefore be either a beast or a god' (ibid: §1253a25). Further, for Aristotle, man's interest in political association is not a conscious choice, but a natural impulse. The *polis* is a political association that fulfills the nature of man, as famously 'man is by nature a political animal' (ibid: 10 §1253a2). As the city is more than its parts, it possesses transcendental value and therefore to be a proper politically active citizen is to be connected to goodness and justice.

Across his writing, Aristotle dedicates a significant portion of his attention to the rule and management of the household (*oikos*). This attends to the situations and domains a man has to manage in his home, these being relationships with slaves, marital relationships and parental relationships. The former, slaves, are for Aristotle animate instruments and thereby property, although Aristotle (ibid: 14 §1254a13) points out that if automation (my word) were possible and a 'shuttle would then weave of itself, and a plectrum would do its own harp-playing' then slaves would not be needed. A slave, then, is someone who is not his or her own property, but that which belongs to another by dint of being an instrument for them. This is justified through differences of kind or category. Aristotle will not admit of a continuum of likeness between master and slave, but rather a difference by goodness that is accounted for in terms of being temperate, brave and just. However, he does not recommend 'wrong exercise of his [the master's] rule' as this is disadvantageous for master and slave alike. This is because the slaves do not belong to themselves but they are property of another, so therefore they are part of another. Consequently, to mistreat the slave is to mistreat one's self. The man's rule over his wife and children differs in type. Where the former type of rule over slaves is statesman-like, the latter is paternal as with a Zeus-like king over its subjects. The male right to rule stems from the 'fact' that the 'male is naturally fitter to command' (ibid: 33 §1259a37), although elsewhere Aristotle comments that while a man should rule a marital relationship and there are matters clearly to be decided by men, 'matters that befit a woman he hands over to her' (2009 [350 BC]: 155 §1160b). In general, however, a woman's lowly existence is a given in Aristotle and there is no suggestion of entry into the polis. It is less a case of being chained or restricted to the *oikos*, but rather ascension is unthinkable so is not discussed. While clearly gender opportunities were very different, it is interesting to note that although women's lives were highly tied to the domestic sphere there were exceptions and a few women moved in the public domain. For example, Aspasia, mistress of Pericles, deeply influenced the political mindset of her partners and husbands. However, her capacity to participate in the public life of the city and be free of legal restraints that confined most women to their homes occurred because she was not born of Athens, but was an educated foreigner from Milieus, an Ionian city (Glenn, 1994).

The boundary between being public and private was not impenetrable either as there was neighborly interest in unusual events going on nearby (be this novel sexual partners, extramarital affairs, homosexual liaisons and other domestic goings-on). Indeed, Plato (2004 [360 BC]: 221 §6.783e) in *The Laws* in a section titled 'Correct Procreation' asserts that marriage and the domestic realm should benefit the *polis*. This involves those recently married producing the 'best and finest children' and newlyweds being policed by women (selected by officials who are presumably male), who should report whether the wives or husbands of child-bearing age are concerning themselves with other activities.[1] If the newlyweds continue not to have children, the female officers are to enter the homes of the young people, and admonish and threaten them with making the case public by means of public notification. However, in general (and there are a few other exceptions in *The Laws*), political life stands in distinct contrast to the private realm (*oikia*) and the family.

To be non-public and private is to be of lowly status, if of any status at all. A more detailed account of the etymological roots of privacy defines it as *privo*, *privare* and *privatum* that suggests a deficient state characterized by privation. *Privo* means to bereave, deprive, rob or strip of something; *privare* means to be free, released or delivered of something; the latter, *privatum*, and its adjective *privatus*, means apart from the State, peculiar to one's self or belonging to an individual (Engelhardt, 2000: 121). Notably in *The Laws*, Plato phrases privacy in negative terms that readily connects with the well-known refrain of 'having something to hide,' also addressing privacy as a threat to the social good. Instead, people for Plato should be well-known one to another and should not be kept in the dark about each other's character. To add to Greek ideas about privacy as a deeply negative social status and as being a state of deprivation, it was also seen as that which hides and covers-up. For Plato, citizens (men) who conceal themselves and are unknown to others will not enjoy respect 'or fill the office he deserves or obtain the legal verdict to which he is entitled. So every citizen of every state should make a particular effort to show that he is straightforward and genuine, not shifty' (2004 [360 BC] 160–161§5.738e). The idea that privacy might have a positive dimension whatsoever does not occur to Plato anywhere in *The Laws*.

Arendt and the rise of the social

Against the background of public/private distinctions in ancient Greece, Arendt (1998 [1958]) makes critical observations about what she sees as the death of contemporary public life and the rise of the social. Her suggestions are best seen as a reaction to mass society, modernity and of domination, rather than forums for debate. For Arendt the political and the social should not be confused, as the social is predicated on modernity and ideas about mass society. This sense of the social over the political is expressed at a number of scales involving economics, bureaucracy, nationwide housekeeping and a totalitarian sense of one-ness over a plurality of views and divergent opinion. Crucially, the rise of the social involves a loss of the public. Arendt laments, in these situations, 'men have become entirely private, that is, they have been deprived of seeing and hearing others, of being seen and being heard by them. They are all imprisoned in the subjectivity of their own singular experience' (1998 [1958]: 58). Where healthy public spheres involve debate, ongoing discussion and disagreements, Arendt's argument and concern are the ways in which uniform views come to be. Related, and possibly of greater concern, is intolerance to other views as the very idea of tolerance involves ongoing scrutiny of the capacity for difference and dissent. Arendt is deeply pessimistic about the possibility of a public sphere in modern society (presumably leaving aside slavery, domestic inequality and other negative aspects of ancient Greek arrangements) as the rise of the social promotes homogeneity over discussion and dissent, and it is here where Arendt begins to share commonalities with Habermas and his call for open debate. Benhabib (1993) on Arendt likewise remarks that the rise of the social over the political involves a transformation of what was once public into a pseudo-space for interaction, and one in which we 'behave' as consumers, producers and city dwellers rather than as civic participants. For Arendt as political life slides into social housekeeping, mass society liquidates both public *and* private realms, and any sense of shelter that we might have had from the world.

Media

The point about the dissolution of public/private distinctions is particularly acute in our own period, not least because of the exponential increase in mediated and networked surveillance. Arendt (1998 [1958]) is oddly silent on the interconnections between the media industry, the rise of the social and the concurrent demise of the public and the private. In Arendt's own time of writing she witnessed both the mass mobilization of totalitarian war machines, the post-World War II spread of capitalism, exponential increase in advertising and the permeation of market forces into the home. Media are strangely absent in her account of forces that shaped the modern world. While much has been written about the influence of mass media in the middle of the twentieth century in regard to persuasion, ideology and corporate and political interests placed above the citizenry (Thompson, 1963; Williams, 1992; Curran, 2002; Briggs and Burke, 2009), we might also read this as a border breakdown between public and private interests. Contemporary media enlargement continues this theme through intensification of networks, increase in feedback technologies and analytics, and the novel techno-social arrangements these contribute to.

We might also highlight the ways in which borders of physical and virtual varieties are highly porous when it comes to the monitoring and surveillance of our behavior and interests. Arendt's dystopian visions readily align with those accounts today that allude to the collapse of border-based arrangements because of the reach of modern computer networks and telecommunications, and the ineffectiveness of borders to prevent what was hitherto private being used for commercial and surveillant purposes (Lyon, 2001; Arvidsson, 2004; Elmer, 2004; Andrejevic, 2007; Hildebrandt and Gutwirth, 2008; McStay, 2011). Less pessimistically, we can also point to a net increase of watching taking place. In recent years the capacity to see, mediate and share goings-on has been extended and intensified so to encompass sousveillant media and dissemination (Mann et al., 2003; Bakir, 2010; Mann, 2013). We should not for a moment lapse into utopianism as many services we use for these activities are privately owned and subject to both commodification and centralized surveillance. Indeed, given Edward Snowden's revelations of mass surveillance of citizens' telecommunications usage by US and UK intelligence agencies in 2013,[2] it is all too easy to trace a path from personal sousveillance and lifesharing to surveillance as traces of our personal lives are operationalized in the name of security. However, undeniable is the mutual nature of this visibility, the bringing about of a more level (but far from equal) playing field, the capacity for institution watching, and the noopolitical and resistant dimension to this generated through networked media (Terranova, 2007). While public/private borders are porous and untenable, dystopian narratives should be avoided because the enhanced milieu of visibility encompasses the breakdown of borders of traditional power-holders and citizens alike.

Conclusion

This chapter has identified ancient Greek accounts of privacy as a key discursive source for contemporary understanding of privacy, possibly only second to liberalism (the subject of the next chapter). Such dualistic and border-based conceptions of privacy continue to permeate both academic and lay thinking. Also locatable in both Plato and Aristotle is mistrust in being unnecessarily private, culminating in Plato's 'shifty citizen,' or that private person today 'who has something to hide.' This chapter developed border-based narratives in reference to Arendtian arguments about the implosion of private and public domains, and the rise of the social. While the omission of media is strange given her interest in power, politics and the lives of citizens within totalitarian life, this is easily filled in. The rise of mass media brought with it nascent expertise in influence, strategic communication, and manipulation of desires and fears thus making public what was private, and the contemporary media environment enhances this. However, we should not wallow in Arendtian dystopian narratives but recognize that mediated sociality also grants opportunity for increased visibility of political machinations, resistance and affirmative noopolitics.

CHAPTER THREE

Liberalism, consent and the problem of seclusion

Alongside border-based conceptions, the domain of philosophy most influential on privacy is liberalism, particularly in relation to ideas about rights, non-interference, freedom, autonomy and consent. Indeed, many liberals employ the phraseology of border-based conceptions with Locke, Mill, Constant and de Tocqueville all accounting for liberalism as involving a minimum area of personal freedom that must not be violated. Predicated on frontiers and the protection of personal domains, manufactured divisions are needed to separate public and private. Indeed, many liberals argue that such need for privacy is integral to being human. However, the idea that citizens might have inalienable rights and that limits might be placed on political power was certainly not the case in early modern societies, and while Ancient Greece has been a useful place to begin our philosophical journey on privacy, it is liberal ideas that directly inform contemporary rights-based conceptions of privacy. Be these debates over the right to be forgotten, or to be left alone, privacy is inherently bound up with liberalism.

Historically rights have been from the state but over recent decades this has increasingly involved corporate actors so to comprise a political-economy nexus (forcefully underlined by the 2013 Edward Snowden leaks). Debates about the constitution of consent, whether it can be implied or explicitly stated; the nature of the relationship between people, governments and businesses; and related questions we today take to be indexical of privacy all have their roots in liberalism.

Without understanding this sometimes contradictory but fertile philosophy we cannot hope to have a proper grasp of the construction and significance of privacy in the ongoing societal discussion of appropriate relations between citizens, state, business and questions that inform international policy-making on information. For example in Europe at the time of writing, the General Data Protection Regulation was proposed to update the 1995 Data Protection Directive. This has seen wrangling and high-levels of lobbying as it proposes measures that provide people with greater control over their data and ensures that businesses that handling data abide by the rules. However, while intended to address the contemporary media and e-commerce environment, many of the sticking-points, topics and questions being fought over by member states and lobbyists are those intimately associated with basic liberal principles: consent, autonomy and non-interference from unwanted others (along with more up-to-date questions about whether anonymity is possible in an age of 'Big Data'). Indeed, a read-through of any Article 29 Working Party document (that advises the European Commission) sees passages, terms and discourses easily locatable in either Kant's account of freedom and autonomy, or Mill's approach to liberty. For these reasons this chapter unpacks liberalism in relation to privacy, paying particular attention to ideas about consent so to better understand the philosophical backdrop that feeds ongoing discussion about consent, and boundaries of the self in relation to governments and corporate actors. However, liberal accounts do not have everything their own way, and as we proceed through later sections of the chapter I will highlight weaknesses and oddities in liberal approaches, particularly in regard to emphasis on the individual over collectives.

Toward self-definition

Liberalism is that domain of political thinking that believes that despite possible risk to the wider social body, people must be accorded personal freedom and the right to privacy. Contemporary liberalism in relation to privacy is driven by concern about mass surveillance, social conformity, tendency towards political homogeneity, governmental misuse of information, human errors and false positives, unintended uses and consequences, fear of future governments, and the fact that surveillant and governmental resources once granted and constructed will never be disassembled—even in politically stable periods, not least because of the boons total informational awareness provides for police and local authorities. The problem for liberals is that governments cannot be relied upon to safely look after such information in a responsible and open fashion so to be properly accountable to the people it governs. Liberalism also underpins broader premises we take for granted, such as: private property, freedom of expression, freedom of contract, limited government to prevent the use of force against its own citizens, and societal infrastructure (roads, courts, defence) needed for the protection of private life. Enlightened deliverance of liberty is predicated on ideas about autonomy, the guarding of 'life-projects of the self,' self-development, exploration of individuality, and developing possibly unusual ideas and practices free from unwanted interference from the state and other actors. In short it involves self-definition and direction, and the possibility of developing a life that runs counter to convention and dominant cultural discourses. Privacy in a liberal context is predicated on the capacity to manage access to one's life—assuming no significant impact on others and that a common political good is still advanced. While this may appear to be common sense, as demonstrated in Chapter 2, other worldviews are possible, and the wish to pursue what we see as valuable in our own way is a modern and somewhat self-involved idea.

We might point out that philosophers and political thinkers placed under the rubric of 'liberal' did not all set out to be liberals (Waldron, 1987). Rather their aims were to presents theories of government, society, ideas about political economy and conceptions of how we might better live together. Only consequently did this wide range of philosophers and political theorists receive the designation 'liberal.' Further, in making their original points, liberals drew on work that was influenced by both socialist and conservative trends. This provides all sorts of uncertainty for those seeking a pure strain of liberalism. Liberalism's origins are not entirely clear with some locating it in Locke on property and liberty, in Hobbes and wars against religion, and others with Mill and his emphasis on freedom from social control (Rössler, 2005). Whatever we take to be the starting point, it is indexical with the Enlightenment (circa 1689 to 1789) and that general sense of self-determinism, respect for reason and, certainly in Kant's (1983 [1784]) account, the privileging of moral enlightenment over raw knowledge. Enlightened political understanding means that we are able to discern what humankind wants; create technical means to satisfy these; engender wisdom, and deliver virtue and happiness; fashion a common understanding of what is good; and establish public space where encounters between reasonable people may take place. Further, in establishing what is private and public, liberalism pertains to protect the former from the latter.

Enlightened and rationalist accounts of morality, while popularly adopted, find most overt early expression in Samuel Clarke (2013 [1728]), who proposed an objective and mathematical-type model of morals, or as close to this as possible. This is not quite a causal or absolutely rigid approach to morality but rather an approach that sees responses to situations as having a 'best fit,' or a 'suitableness' to them. By being rational the fit of a response to a situation is plain to an intelligent being. Rather than morality as that which is blindly followed, for Clarke people can discern ethical truths. In contrast to the determinism of Hobbes and Spinoza (to whom Clarke was responding in his book) people are capable of being the first and prime cause, and are not simply held to the unfolding of determinism. The capacity for free will is a fundamental point because not every effect for Clarke is the product of an external cause, and it is this predicate that allows Clarke to build an account for the possibility of liberty (in opposition to necessity and fate). It is to grant the power of choice. Not all of a liberal tendency agreed with such logical approaches to morality. Hume (1965 [1739]) for example seeks to remove both religion and reason from morality. While no romantic of any sort and a clearly able logician, morality for Hume is not to be found in rational decisions, or the working out of the 'best fit,' but in the act itself. Passions or actions are neither true nor false, nor can they be counterclaimed against. Reason and moral beliefs are different and if they are to be equated with geometry or algebra, then this means they should have analytical power in regard to providing us certainty on our decision-making. An engineer for example in designing a bridge will be heavily reliant on the certainty of angles and measurements to address load-bearing. Does a moral framework have the same degree of everlasting reliability? For Hume they are different categories and morals are not to be equated with numbers, quantity or degree of quality.

Between self and society

The idea of rights as naturally attributable to people is best connected with Hugo Grotius (2001 [1625]) and his observation that norms and rights exist, even if it were discovered that there is no God, or that he is not interested in our affairs. It is through Grotius that 'human rights' and the capacity for individuals to pursue their own good becomes a central pillar of society. Best expressed in *On the Law of War and Peace*, his account of natural law contributed to the framing of philosophical and political developments of the seventeenth and eighteenth centuries. His approach recognizes the need and value of community, but somewhat uniquely individual rights come to a hitherto unseen degree of prominence. This is because for Grotius rights are natural qualities that ground and organize community law, rather than the other way around of laws determining rights. Grotius notes however that they may be traded, they might not result in the protection we are due, and they may not be recognized by others. Nevertheless, in Grotius we can locate the inception of *rights as property* that remains prevalent today, or that sense of indignation that allows us to feel aggrieved when our rights have been contravened in some way.

Hobbes (1985 [1651]) also (but differently) displayed an interest in early forms of liberty discussing it in terms of absence of impediment. To guarantee a functioning society and avoid a collapse back into a 'state of nature' or brutish self-interest, Hobbes (1998 [1668/1642]) proposes a common and central power for which we put aside a right to all things in favor of common interest, governance and mutual contracts with fellow humans. Where in the hypothetical state of nature we have rights to all things (rendering property and privacy meaningless), the setting up of magistrates and authority requires the relinquishing of rights. After all, for Hobbes at least, before civil society no one had 'proper rights' and 'all things were *common* to all men' (1998 [1642]: 249 [emphasis in original]). The founding of liberty means that rights are given up for the establishing of peace and the avoidance of perpetual war. For Hobbes this rational seeking of peace is a natural law but, by extension, it also follows that it is a natural law *'that the right of all men to all things ought not to be retained; but that some rights ought to be transferred or relinquished'* (ibid: 123 [emphasis in original]). This requires consent of the people to be governed. In exchange, people obtain protection, providing they obey the law. Having a somewhat pessimistic view of people, Hobbes argued there can be no universal benevolence and that we will not show equivalent care and love for other people when they have no connection to us. Moreover, if a large amount of people are to live together, there must be rules and if there are to be rules, there also needs to be a common power to enforce this. To progress to such a commonwealth, some pre-social contract natural rights (involving freedom from contractual obligation) need to be given up for the wider benefit of social stability. In giving over some of these rights, this gives others (those who will govern) rights by means of contract.

Here we lay aside the right to act with only self-interest, and instead grant others authority and power over us. This is not simply an abstract agreement, but also involves means of enforcement. Being more powerful than any person in the state of nature, an artificial person, Hobbes' (1985 [1651]) *Leviathan*, is able to mete out punishment to those who transgress the contract. This then is a coercive power characterized by might (as in nature) and acquired right (contract). The primary task of the sovereign then is to keep us safe and offer protection, although Hobbes recognizes that governance requires more than this—it requires we be happy too, although this has to be balanced with security needs. While Hobbes can make for mildly depressing reading, particularly given his view of humankind, he does observe in *De Cive* (The Citizen) that safety cannot come at any cost or be the sole concern of life in society, but a society should also 'furnish their subjects abundantly, not only with the good things belonging to life, but also with those which advance to delectation' (ibid: 259). This particular point is a fundamental one as it underpins and explains continuing tensions between protection and safety, and privacy and liberty.

Locke (2005 [1690]) portrays how each person absolutely possesses themself, is able to extend these property rights by sweat and labor, and that what is acquired by the self is private property. In Locke we find deep roots on privacy as a natural right and the connection of privacy to property. Some origins of modern ideas about consent are also found in Locke, as in expounding his theory of government he defines legitimate governance as that which has the consent of the people (and that which does not, may be overthrown). Locke's (2005 [1690]) *Second Treatise on Government* also addresses the state of nature theme, but sees it differently to Hobbes in that for Locke a pre-political natural state is one characterized by equality, although without a state this remains fraught with danger as there are no established norms or judiciary to take recourse to. With such an emphasis on freedom and lack of accountability to authority, why would anyone make a pact to give up such freedom for political union and civil society? The answer for Locke is in privacy and property, or to protect private property. Rousseau's (2008 [1762]: 27 [emphasis in original]) take on the social contract is similar and by entering into it, 'Man loses by the social contract his *natural* liberty, and an unlimited right to all which tempts him, and which he can obtain; in return he acquires *civil* liberty, and proprietorship of all he possess.' However, Rousseau has a very different take on privacy. While Hobbes and Locke see the social contract as a means of ensuring stability to pursue private interests, albeit with certain freedoms given up, Rousseau sees privacy in negative terms. A quintessential author of the Enlightenment, Rousseau's (2004 [1755]) *Discourse on the Origin and Foundations of Inequality Among Men* sees privacy as responsible for the fall from grace within a state of nature—specifically the coming to be of private property. This comes to be because of talent differentials and who has the most strength for labor. While two might work equally, a natural inequality unfolds. This brings about rivalry and subsequently envy and greed. Differences between people become apparent and more permanent in its effects in regard to fortunes and employment of others (and thereby greed, competition and related ungraceful values).

Positive/negative liberty

For Berlin (2006 [1958]) liberty and freedom are near enough to be synonymous. Berlin's essay, *Two Concepts of Liberty*, reports two senses of freedom and liberty—negative and positive. The former is defined in terms of the absence of coercion, obstacles, barriers, constraints or interference from others; the latter is more active involving taking control of one's life and realizing fundamental purposes, be this for an individual or a group. Put otherwise, one has to do with obstruction and the other self-determination. Negative ideals involve liberty *from* unwanted actions of others. Whether this involves natural or civil rights, an imperative, agreed rights predicated on utility, a social contract, or any other means of setting up norms, the conclusion for Berlin is the same—liberty has to do with avoidance of interference from others. This is less about seclusion or being alone, but according rights to individuals so for the fullness of individual human potentiality to be recognized and respected. The consequence then is a gravitational shift of emphasis from privileging structure and process to acknowledgement of richness of human being, and again that sense of people as prime causes. Berlin draws on Mill's (1962 [1859]) *On Liberty* (discussed in the following chapter) to argue the value and need for liberty. Here Berlin (along with Mill) maintains that without the capacity to live as one wishes in areas of life that do not affect others, free ideas will not come to be thus retarding the scope for innovation, originality or moral courage. Berlin's positive sense of freedom and liberty involves 'freedom to' rather than 'freedom from' (as with negative conceptions). The longer description of positive freedom is worth quoting:

> I wish my life and decisions to depend on myself, not on external forces of whatever kind. I wish to be an instrument of my own, not of other men's, acts of will. I wish to be a subject, not an object; to be moved by reasons, by conscious purposes, which are my own, not by causes which affect me, as it were, from outside. I wish to be somebody, not nobody; a doer—deciding, not being decided for, self-directed and not acted upon by external nature or by other me as if I were a thing, or an animal, or a slave incapable of playing a human role, that is, of conceiving goals and policies of my own and realizing them. This is at least part of what I mean when I say that I am rational, and that it is my reason that distinguishes me as a human being from the rest of the world. I wish, above all, to be conscious of myself as a thinking, willing, active being, bearing responsibility for my choices and able to explain them by reference to my own ideas and purposes. I feel free to the degree that I believe this to be true, and enslaved to the degree that I am made to realize that it is not. (2006 [1958]: 44)

Enslavement for Berlin (reduction of positive freedom) is determinism by another name. Elsewhere Berlin remarks that if the full consequences of determinism were taken forward, 'the basic terms and ideas that this realization would call for would be greater and more upsetting than the majority of contemporary determinists seem to realize' (1969: xiii). While not explicit, we can presume these feared outcomes revolve around what happens when responsibility and agency are diminished. This can be unpacked in terms of the consequences of handing over decision-making powers to others because they are in some way better placed to handle our 'best interests.'

Conflict and paternalism

Berlin gears-up his discussion of rationality to question the confrontation of our empirical selves (our flawed everyday selves that acts on impulse) with our rational self that plays the longer game, and with which our empirical selves must be made to coincide. This is a recurring theme within privacy discourse in that those of us interested in privacy seek to correct behavior, educate, point out the error of others' ways and make people aware of what we believe to be the implications of inaction or non-awareness. We preach a normative account of privacy, and seek ways of enlightening (educating) others into better privacy behavior so to accord with what they would rationally want if they truly understood. This is paternalism, or that which if taken to conclusion means we will forcibly stop someone doing something if it is not in their long-term interests. Can paternalism however turn into authoritarianism? That is, if there is *a right way of doing things* (i.e. *one* best way) as espoused by areas of reason-based Enlightenment doctrine, should the rest of us unwise and unknowing folk be dragged into line? Post-Kantian thought (Johann Fichte and Auguste Comte) tends towards this, although Kant's emphasis on individualism stops himself from this tendency. Berlin's wider point is not that positive liberty must be predicated on autonomy and not on rational (mono-logical) notions of freedom, for this gives rise to the capacity for others to 'act in our interest' and therefore totalitarian worldviews.

Deontology

Absolutist premises and deontological propositions typically include rights, responsibilities, obligations, duties, privileges, entitlements, penalties, authorizations and permissions (Searle, 1996). Prior to the Enlightenment, the idea that people might have rights by sole dint of being human did not exist. While initially a divine authority provided rights, the idea has persisted so for the idea of human rights to continue without need for religious belief. In Kantian (1991 [1785]) terms this involves the categorical imperative, the maintenance of morality as a universal law, duties and obligations, and the coming into being of these because of reason. The power of Kant's imperative is that it underpins all other ethical propositions (for example, not murdering, stealing, harming and so on) and each flows from the imperative to act in such a way that we will it to be universal law. The idea then is that the subject through their own reasoning will arrive at an objective law not contingent upon desire, preference, circumstance or even personal principles (they should be valid for all, not just the subject). The key reason for the imperative's persisting influence is that moral action comes to be by means of acting in a way that our intentions and behavior would become moral law. The radical part of this is that no outside agent determines what that law might be and the only moral requisite is that what we will would be law. Kant's rational agent is a vital component for liberal ethical theory because it is by means of rationality that the subject will arrive *autonomously* at principles they believe to be objective for all. The subject, then, is the author of his or her own principles but these are arrived at by logic, rationality and without recourse to the vagaries of circumstance. This means not being swayed by context, wishes, desires, interests, outcomes or what Kant calls 'passions' so to stay focused on what is objectively proper. Unobjective factors involve what Kant designates as *heteronomy* and non-moral imperatives. In these situations the will is compromised and bound to an objective or end that is not its own. Instead in Kant's view we should act objectively, universally, for all people at all times, and not just for individuals in specific contexts. For Kant a right is

> ... the limitation of each person's freedom so that it is compatible with the freedom of everyone, insofar as this is possible in accord with a universal law; and *public right* is the totality [*Inbegriff*] of *external laws* that make such a thoroughgoing compatibility possible. (1983 [1793]: 72 [emphasis in original])

This means that autonomy-led accounts are not as selfish as they are sometimes accused of being as they are not simply private but, more importantly, public. Kant's understanding also involves *a priori* norms of freedom, equality and independence as a citizen. A key point about such deontological premises is that equality-based rights to autonomy cannot be traded or terminated under any circumstance. This has consequences for recent market-based ideas about privacy (and Grotius' point about the potential to trade rights). While much of what Kant says on autonomy we take today as given, there are glaring inconsistencies in Kant worth highlighting— not least in regard to his treatment of women. Reminiscent of the discussion of ancient Greece in the previous chapter, Kant on citizenship remarks:

> Any person who has the right to vote [*Stimmrecht*] on this legislation is called a citizen (*citoyen*, i.e. citizen of the nation, not citizen of a town, *bourgeoisie*). The only quality necessary for being a citizen, other than the *natural* one (that he is neither a child nor a women), is that he *be his own master* (*sui iuris*), ... (1983 [1793]: 76 [emphasis in original])

Not all Enlightenment commentators shared Kant's paradoxical views. Mill for example found sex-based discrimination dubious, inefficient and therefore unjust. Women for Mill have as much right to liberty and self-development as men, and sexism for Mill is an irrational interference with personal initiative and laissez-faire. (Mill's support was not only theoretical, as he petitioned the English Parliament for universal suffrage.) While in many ways a champion of individual freedom, Kant also has what today are difficult views on the management of public and private domains. For example Kant takes a hardline on the need for respect of administrations. For Kant, the running of a commonwealth is not to generate happiness but to ensure its survival. More specifically, this survival involves the continuation of deontological rights and in Kant's political philosophy, we must never oppose legislators however much pain they cause or however terrible their decisions for this would destroy the civil constitution. To quote:

> From this it follows that all resistance to the supreme legislative power, all incitements of subjects actively to express discontent, all revolt that breaks forth into rebellion, is the highest and most punishable crime in a commonwealth, for it destroys its foundation. And this prohibition is *absolute*, so that even if that power or its agent, the nation's leader, may have broken the original contract, thereby forfeiting in the subject's eyes the right to be legislator, since he has authorized the government to proceed in a brutal (tyrannical) fashion, the citizen is nonetheless not to resist him in any way whatsoever. (1983 [1793]: 79 [emphasis in original])

His reasoning for this deeply unpalatable argument is that rebellion would introduce lawlessness so that aforementioned deontic rights would lose their sanction. Aware of contemporaries who would plead on behalf of the people, he concludes that we should not substitute the principle of happiness for the principle of rights. This is by definition the deepest of conservative views as it is founded on the perceived need to maintain the status quo at all costs, and avoid bifurcatory and transformational situations.

Consent and autonomy

The liberal idea of consent is founded on an autonomous self capable of making decisions free from excessive constraint and external pressure. The premise of autonomy as a moral matter is fully attributable to Kant (in his more positive writing), although the premise of self-governance has a longer history. Kant's key argument on morality is that people impose laws on themselves and in doing so provide themselves with a motive to obey. This is expressed well in this passage: '*Enlightenment is a man's emergence from his self-imposed immaturity. Immaturity* is the inability to use one's understanding without guidance from another' (1983 [1784]: 41 [same emphasis as in original]). Indeed, Kant opens his famous essay *What is Enlightenment* with *Sapere Aude!* or 'Have courage to use your own understanding!'. For understanding to flourish, freedom is required. More specifically this idea of freedom is that of a public sort—that which underpins the free exchange of ideas, reason and debate. It is in opposition to commandments to obey. This renewed and enlightened self is that which rejects religion (or 'self-imposed immaturity') as pernicious and disgraceful. It is also that approach that requires a government to treat people with dignity and as more than machines. Autonomy, self-guidance and the ability to choose is for Kant the basis of a moral life and it is a key social imperative to participate in this, to protect it, and provide circumstances for it to develop elsewhere. This is in stern contrast to any philosophy that sees morality as obedience. The morality as obedience approach has two discernable factors: one, as created beings we should display deference and gratitude; two, our moral capacities are not up to the job and we are either too weak-willed or strongly driven by desires to live autonomously (Schneewind, 1998). Kant rejected both and the gravitational shift towards selfhood, freedom and autonomy is heady stuff. Kant does not reject God altogether but makes the eyebrow raising claim, and what in the 1800s would have been a jaw-dropping assertion, that God and people can live in a single moral community only if we have similar entitlements to constitute the rights we are going to be required to obey. We are able to occupy the same level as God not because of any deep understanding of eternal moral truths, but because we can both make and live by truths (or moral laws) arrived at autonomously. This is an assertion of both will and freedom, and an outcome of the categorical imperative.

A key problem of accounts predicated on autonomy is that it is difficult to place within the context of everyday life and the privacy events we find ourselves in. As a principle it is right, fine and noble, but from the point of view of making policy that applies in all situations it is difficult to implement. This is because in a liberal context the self is seen as a stable rational entity able to make clear choices, but lessons from social constructionism, and those areas of sociology that have sought to overcome questions of structure versus agency, show this to be problematic (Giddens, 1984). Also, with acknowledgement of the contingency of the field of knowledge that informs decision-making processes, the scope of the autonomous self is reduced somewhat as constraints are placed at the time of choice (meaning that limited options are not real choices). Moreover, in making privacy decisions, short-term and long-term goals might conflict so for us to require upfront assistance to make the better longer-term decision.

Tacit vs. informed consent

Inextricably connected to choice and autonomy is consent. This is established by multiple discourses within liberalism from social contract theorists onwards who highlight the need to recognize the relationship between citizenry, will, agency and consent as an expression of autonomy. Tacit consent is a point addressed in full by Locke (2005 [1690]) who describes that it involves enjoying the benefits of land, lodging or carriage within the territory of a given government or domain. People in this account are free to leave, or in more contemporary parlance, opt-out. Tacit consent is silently given and unless actively dissented against, consent is assumed. There are arguments to be made against this, not least that we are often unaware of what is taking place and, as Hume (2005 [1753]) reminds us, this is akin to being carried on to a ship asleep and being said to consent to the dominion of the shipmaster. The question then is whether being in a domain constitutes *actual* consent? More generally, does consent require an expressive element or can it be passively given?

While questions of philosophy, they are also of pivotal importance for understanding the interrelationships between technology, media, business, citizenry, legal and regulatory bodies, and governments (also see McStay, 2013). As mentioned in the introduction to this chapter, at the time of writing Europe is reconsidering its data protection laws. Historically the European Commission implements directives that are adopted into local law in each of the European Union's 28 member states, but this update involves the introduction of a new single data protection law. The idea is to bring an end to the fragmented regime so for all members to be able to work more clearly within one single law. However, many member states (the UK in particular) and corporate actors complained that the proposed regulation is too prescriptive. Among a range of sticking points is consent. The European Commission seeks a situation where citizens have to provide 'clear affirmative action' that involves explicit consent to the processing of data. Businesses, and many Ministers, prefer a form of consent that is 'unambiguous.' The language that defines this is unclear, but the former involves consent that is given in either oral or written terms. It is to be freely given and the citizen is to receive sufficient information to understand the scope and consequences of consent. Further, consent must be specific leaving no doubts as to whether it was given or not. In contrast, the recommendation on unambiguous consent does not state whether or not an opt-in action is necessary and if data collection is the default situation. While dry material, the implications are significant as much can hang on legal technicalities and definitions of consent (McStay, 2013). While admittedly on paper the opt-in (upfront stated acceptance) and opt-out (data collection as default) debate may appear a semantic difference, the ramifications for the technology and communications industries are significant. Moreover, the entire situation is utterly saturated with liberally framed values and questions. Beyond these opt-in/out considerations, there are even larger implications for how we conceptualize and value consent as a liberal norm within society. Do we place greater importance on voluntariness, privacy and self-determination; or do we privilege the unfettered flow of information so to facilitate business and commerce?

Problematizing privacy

Liberal accounts of how society should best function involve autonomy and the freedom to do what one wishes as long as it does not negatively impact on others.

Indeed, privacy is nigh-on synonymous with liberal ideals, particularly in relation to J.S. Mill's (1962 [1859]) *On Liberty* and the premise that one is not accountable to others as long as these interests do not concern others. This is arguably the most influential contribution from philosophy to privacy, even surpassing Greek dualistic premises and the shifty citizen accounted for earlier. However, as pointed out, despite the caveat of avoiding negative impact on others, radical autonomy involves theoretical excess because even the most anti-social of us exist within complex arrangements of others. There are two dimensions to this: first, in regard to questions of social structuration and the possibility of autonomous decision-making; second, a less intellectual observation about human inter-dependence. While privacy is important, this should be tempered with a need for private accountability in terms of law, morality and social norms. It is to balance the interest of safety, health, security, efficiency and other people, with liberal values of privacy. Critically, a degree of accountability protects those people in society less able to protect themselves. Allen (2003) suggests that accountability dignifies and by this she means that a society that makes individuals accountable also credits them with intelligence, agency and common respect (unlike animals). On living with others, Schoeman criticizes Mill and his followers for not paying adequate attention to privacy (and liberty) in terms of collectives and groups. Where many liberal accounts focus on individuals and their need to avoid unwanted external pressure, Schoeman is interested in contexts, the possibility of joining alternative groups, and the means by which people should have opportunities 'to pursue with others significant ends without enduring unfair or unreasonable social sacrifices' (2008 [1992]: 193). This is to suggest that groups, as well as individuals, have rights to privacy. As argued throughout this book, privacy is best seen in terms of protocol and the modulation of connection with others, and not simply seclusion.

Conclusion

The aim of this chapter has been two-fold. It first sought to offer understanding of liberal philosophies paying particular attention to the subject as prime cause, contract theory and the observation that living in commonwealths within a judicial system and under state protection involves giving up a degree of freedom. The chapter progressed to account for the theoretical roots of autonomy that are clearly linked to Kant, the categorical imperative, the avoidance of heteronomy and the willing subject able to make clear decisions. This allowed us to advance into the second part of this chapter so to assess consent and the ways in which contemporary technical legal questions about the constitution of consent, new media, cookie use and data processing and analysis have origins directly locatable in liberal philosophy. While legal specifics are both difficult and perhaps unexciting they are of real consequence, and we might also see such debate about the constitution of consent as philosophy in action, and as deeply affecting, shaping and organizing our modern world.

However, there are criticisms to be made of the liberal worldview, not least its deep emphasis on autonomy and excessive promotion of individualism. The anti-paternalist aspect is also difficult to tally with contemporary life, particularly given the overwhelming nature of informational privacy, the number of privacy decisions we are asked to make and the difficulty in understanding the terms of those decisions. Left to ourselves our online lives would be utterly unmanageable, and on this point I agree with Allen (2011) who suggests that paternalistic privacy laws should not be dismissed out of hand, particularly if we agree that privacy is a foundational human good.[1] The importance of the capacity for autonomy is retained throughout this book, but the emphasis on seclusion and being alone is rejected because privacy involves openness as well as being reserved. Also, we live in communities and fields of influence from which it is impossible to be entirely independent of. There are two ways of seeing this: 1) by means of the Archimedean principle and the impossibility of being objective when we are inside a situation; and 2) how does a moral compass find its bearings if morality is grounded in the self? In contrast to excessively self-focused narratives, this book progresses on the basis that morality is grounded in culture and that the aim of culture is living in healthy community with others. This is a markedly different account from those predicated on seclusion, the right to be left alone, solitude and those that privilege the self over collective interest. In regard to the protocol view of privacy being promoted in this book, the emphasis on the atomistic individual equates to privacy as seclusion. I disagree with this because as argued, privacy is *always* about connections and relationships. This is to frame privacy in terms of *everyday mutuality*, and maintain the best of liberalism's emphasis on dignity and respect, without its excessive interest in seclusion.

CHAPTER FOUR

Utilitarianism, radical transparency and moral truffles

Utilitarianism has a mixed relationship with privacy. While ostensibly liberal in its approach, there is a latent tendency within certain strains of utilitarianism towards not only enlightenment norms of transparency and calculative rationality, but also what I term *radical transparency*. This chapter accounts for the relationship between utilitarianism and privacy, and assesses the notion of radical transparency. This is developed in reference to Richard Posner, the Benthamite judge and economist, who argues that privacy is tantamount to market inefficiency and hinders that which delivers net benefits to society. These discourses are increasingly prevalent today, not least in the legal wrangling alluded to in the previous chapter on constructions and implementation of consent.

First principles: Cumberland to Bentham

Utilitarianism is important for privacy because of the ways utilitarians think about rights. Rights in this arrangement have a more expedient character than in the previous chapter. Bentham's championing of the principle of utility characterizes this well, particularly in regard to his support of doctrine that sees two masters or poles along a continuum by which we should navigate: pain and pleasure. The principle of utility is a norm that recommends that individuals and governments should promote happiness in respect to the community at large. While associated with Bentham and subsequently Mill, Mary Warnock points out in her introduction to Mill's (1962 [1859]) *Utilitarianism* that its key principle of 'the greatest happiness for the greatest number' does not originate in Bentham or Mill himself. Warnock instead locates this in Joseph Priestley and his pamphlet *Essay on Government* in 1768. We might also look to Bentham's interest in Claude Helvetius and Cesare Beccaria for earlier accounts of the principle of utility (greatest happiness for the greatest number) and the premise that virtue equates to the desire for general happiness. The principle of utility is for Beccaria the foundation of human justice and the argument that the 'public spirit… is influenced by general principles, and from facts deduces general rules of utility to the greatest number' (1767: 93). Certainly in Beccaria's case, his own interest in utility derives from Francis Hutcheson (2013 [1725]) who also employed 'the greatest happiness for the greatest number' in his account of moral ends and the promotion of the general welfare of mankind. In the search for origins of utility, we might even go back further to Richard Cumberland who in 1672 by means of *A Treatise of the Laws of Nature* attempted a full-scale assault on Hobbes's *Leviathan*. Here Cumberland argued that experience tells us that it is best to work together for the greatest possible happiness, quite possibly making him the first political utilitarian (Cumberland, 2005 [1672]; also see Schneewind, 1998). Beyond the vague outline of the principle of utility, he also emphasizes its quantitative dimension, the reliance on reason, and the process of moral judgment by calculation.

Regardless of origin, utilitarian sentiments motivated Bentham into action and writing, with Bentham (2000 [1781]) also founding his definition of utility as the obligation towards happiness. This sees a rejection of natural laws and of original contracts (some of which are discussed in Chapter 3) in favor of a view of morality predicated on utility, calculation and the promotion of happiness. Indeed, famously, Bentham (1843 [1792]: 501) called the idea of natural laws and rights 'nonsense upon stilts' seeing these as fictitious and rhetorical devices. Real rights for Bentham are legal rights and these are predicated on what is advantageous to society. This then is to challenge the idea that rights are inalienable and possessed by dint of being human. The happiness principle for Bentham is not a perfect one but he defends and argues for it on the basis that a decisional principle is required and no other is available. He admits that its value cannot be proved and that it is not self-evident, but points out that it is a principle that people can be motivated to follow. While utilitarianism might be accused of operating by an absolute and deontic norm, Bentham would argue that it is not a metaphysical solution but a practical one.

Specifically, utilitarianism involves a tendency that facilitates the happiness of a community rather than diminishes it. Pain and pleasure in the Benthamite worldview are the two principles that decide utility, and it is calculation by these terms that determines what ought to be done. This applies to choices about and for the self, as well as those decisions taken by governments. The key task for the individual and legislator is to ensure, where possible, avoidance of pain and promotion of pleasure as an end in itself. It is noteworthy too that pain avoidance is to be found throughout the natural world, as well as in political theorizing. Bentham, along with those before him, saw pain and pleasure as meta-values that could be calculated and he attempted to shift these from being simply personal and experiential values to those capable of being public and verifiable. For Bentham, sub-values by which to assess pain and pleasure include intensity, duration, certainty/uncertainty, propinquity (closeness)/remoteness, fecundity, and extent (who is affected by a decision). Bentham suggests that these be broken down into pleasure on the one side and pain on the other, and whichever course of action on balance is more pleasurable is the one to be taken. Bentham's idea of pleasure is wide-ranging and, although based on gratification, encompasses: senses, wealth, skill, amity, good name, power, piety, benevolence, malevolence, memory, imagination, expectation, association and relief (with many of these having the potential to cause pain too).

Mill: A friend of privacy

Mill cites Bentham as a clear influence but equally posits Samuel Coleridge as a seminal mind of his age. While ostensibly agreeing with Bentham about the greatest happiness principle, Mill remarks that happiness expressed in terms of utility is of little use in regard to human conduct. Mill offers a number of correctives and in *Utilitarianism* he remarks 'It is better to be a human being dissatisfied than a pig satisfied; better to be Socrates dissatisfied than a fool satisfied' (1962 [1859]: 260). Mill encompasses within his account of utility the well-being of humankind, freedom of opinion and expression, and other romantic themes that do not feature in Bentham.[1] Indeed, freedom does not come up for discussion in Bentham and it is on this point that Mill makes himself a friend of privacy. Mill's (1962 [1838]) essay, *Bentham*, is praising, but highlights shortcomings in Bentham in regard to the narrow definition of utility. Mill instead seeks an expanded view to steer conceptions of utility away from self-interest, towards ideas about freedom and the pursuit of 'good' on condition that we do not hinder others in their own endeavors. Other factors raised by Mill include the pursuit of spiritual perfection as an end, of self-shaping character towards excellence and of the existence of conscience. For Bentham the self is a discernable and knowable entity, but for Mill it is more open-ended. Mill is relentless in his critique of Bentham but offers him a way out by defining him as a philosopher of business in regard to organizing and regulating social arrangements. So long as Bentham does not encroach upon discussion of morality, Mill is supportive, although he states Bentham 'committed the mistake of supposing that the business part of human affairs was the whole of them' (1962 [1838]: 106).

On Liberty

A point of difference of Mill from other liberals is that Mill eschews deontology and abstract rights, preferring utility. Unlike Bentham however, Mill's utility is 'grounded on the permanent interests of a man as a progressive being' (1962 [1859]: 136). In *Essay on Liberty* Mill offers what is perhaps the longest-standing account of liberalism and the clearest delineation of its principles, particularly in regard to the need for individual liberty. For Mill pleasure is not the aim or end of things, but the goal instead is the creation of conditions that provide a person with freedom to choose and to be left alone if that is what one seeks. Despite Mill being inextricable from utilitarianism, he rebelled against key areas of Benthamite doctrine, particularly in regard to early utilitarianism that saw pleasure and pain as equations to be totted-up and acted upon in some regard. Mill's form of utilitarianism is less about calculating modes of rationality, but rather is inspired by ideals of liberty and justice—again defined in term of right to self-choice. This sits in some contrast to both Bentham and more recent Benthamite commentators on privacy such as Richard Posner who is discussed towards the end of this chapter.

On tensions between authority and wider society, Mill is interested in 'the nature and limits of the power that can be exercised by society over the individual' (1962 [1859]: 126). This power for Mill should only be exercised if a person negatively affects the interests of another. This may appear excessively individualistic and self-serving but Mill's point is that the basis of a healthy social corpus is predicated on free exercise of the will and unfettering of higher critical faculties. Society for Mill works best by each individual knowing the general rules of conduct so we know what we can expect of one another. Also, absence of negative impact on others aside, we are free to pursue our own ways of being. Mill points out that the original meaning of liberty is the limitation of power by means of the establishing of rights and liberties, and also the establishing of constitutional checks to power. A reason to limit power of government, even when conceived in terms of self-governance, is that it tends to be the case that the people who exercise this power are not the same as those over whom it is exercised. Similarly, for Mill 'the will of the people' frequently involves the most active part of the people that are able to convince us theirs is a dominant view. There is discursive subtlety in Mill as he seeks to protect against 'the tyranny of the prevailing opinion and feeling' and the ways in which people in positions of power seek to impose their own ideas and set agendas about conduct and areas of life that for Mill are beyond the interests of the state (ibid: 130). Liberty then is the establishing of the balance between social control and the individual. However, there are wrongs and intellectual crimes in Mill to be righted, particularly over who is ascribed the full rights of a sovereign individual. He discounts children and young people, those unable to care for themselves and: 'For the same reason we may leave out of consideration those backward states of society in which the race itself may be considered as in its nonage' (ibid: 135–136). He also adds, 'Despotism is a legitimate mode of government in dealing with barbarians, provided the end be their improvement, and the means justified by actually effecting that end' (ibid: 136). However, the principle of liberty transcends Mill's shortcomings. As with discussion of Berlin in the past chapter, we might prefer to see Mill's account in terms of freedom to rather than freedom from. This is an important point for privacy discussion as it ensures emphasis is placed on a positive assertion of selfhood and we do not fall into the *seclusion trap*, i.e. a more depressing conception of privacy that mistakes it for solitude and absence of others. Privacy is less about affording 'the right to be alone' (the seclusion trap), but rather privacy is a more positive assertion of how we maintain and manage relationships without interference. This connects to the capacity to control and to determine rather than be determined.

The harm principle

Informational privacy is frequently associated with paternalism and those processes of removing choices from others for what we deem to be their self-interest. This might involve attempts to prohibit or restrict use of social networks due to the intensity of data mining by both known and possibly unknown actors (for example commercial third-parties or security services). On paternalism (or acting for the good of another without their consent and treating them as a father might regard a child), Mill was against this, with exceptions of education and labor legislation. Kant (1983 [1793]) was similarly scathing of paternalism addressing it head-on stating that it fosters immaturity, passivity and relationships with governments that bring about despotism. Although in some cases we appear to act against our own interests, for Mill paternalism invokes an unacceptable trade-off for liberty. There are a number of other limited situations when paternalism is acceptable, most notably in regard to harm to others, and when actions are seen to be interfering and limiting the liberty of others. On consensual paternalism itself, this invokes situations where choice is removed and this for Mill equates to being treated like a child, or as someone weak in the face of influence that needs choices to be removed.

As a principle, for Mill the agent should be able to decide for himself or herself, even if the act is detrimental. Indeed, it is because of Mill's emphasis on choice, self-determination, and the discovering and emerging subject, that he has outlasted the relevance of other commentators who loomed larger in his time. As Berlin (1969: 192) on Mill points out, it is 'his passionate belief that men are made human by their capacity for choice—choice of evil and good equally.' Liberty for both is absence of coercion and the right for people of contrasting persuasions to pursue their own interests. Moreover, the premise that we might all agree was anathema to Mill who abhorred the idea 'of the human pack in full cry against the victim' (ibid: 193). Instead he defended the rights of heretics and dissidents alike (in life as well as in philosophy). Similarly for Mill, surveillance of thoughts, views and beliefs ends up reducing desirable differences to a uniformity of thoughts, dealings and actions. This is at the expense of cultivating what is individual, unique, diverse, rich, animated and of value. It is notable too that this fusion point of romanticism and rationalism moves Mill some philosophical distance from Bentham and early utilitarians accounted for earlier.

Criticisms of Mill and the emphasis on autonomy

Mill over-emphasizes rationality and individuals' capacity to make decisions free from influence of things, people and circumstances we find ourselves in. We are also more deeply imbricated in fields of historico-cultural shaping than Mill's account suggests. Moreover, as highlighted in the previous chapter, the positing of an Archimedean position is deeply problematic. Further, in relation to consent and decision-making online for example, can we free ourselves of behavioral cues, nudges, heuristics, the need for shortcuts and those factors that contribute to decision-making—particularly those of only mild short-term consequence? This then invokes questions of competence and the extent to which we are able to live up to the rationalist ideal of unfettered reasoning. This is a well-known observation, but to this we might add that the liberal premise of unconstrained agency serves the needs of those who wish to make use of our mediated connections with others. This is a key argument for this chapter because Mill's overreach on the role of reason and autonomy means that the very strategy that was meant to lead to our liberation from social pressure is actually that which leads to our vulnerability. This is particularly so if institutions are built around the premise of *oppressive autonomy*, or the proposition and requirement that the self should be capable of responding rationally and in our own best interests whatever the situation. This is because Mill's thinking assumes we can withstand downward pressure to properly and fully decide in every privacy instance we face. The faith in independent reason gives rise to blindness to historical, social, cultural, environmental and material influence. Further, our behavior is not solely dictated by reason, but rather we get annoyed, we need support, give support, enjoy happiness, care and from time to time behave contemptuously towards others. None of us are immune from these sways that impact on our behavior. Even given his romantic sympathies, Mill depicts an impossible reason-based caricature, an archetype or possibly an idol of sorts. From the point of view of assessing behavior in society (and privacy decisions therein) a more encompassing approach is required than that which relies on autonomy and reason because: a) dimensions other than reason play a role in everyone's decision-making; b) external and environmental factors play a greater role than subject-focused accounts admit; c) our rights to autonomy may be used against us.

The key question to those persuaded by liberty-as-autonomy arguments is how do we step outside of ourselves to make these unimpeded and free decisions? How does this Archimedean point come to be? How can we objectively analyze a world we are part of? What of the anthropic principle where what is found by an observer must be consonant with the capacities of the observer? Are decisions made with no dependence or reference to others, wider cultural discourses, circumstances and histories? This not only involves factors we can see, recognize and isolate (as with dominant social discourses prevalent in the media for example), but what about situations where we are not aware of how our actions and decisions are being steered? Critically, how can we bracket-out influences we are not even aware of? Context, culture, environment, options available, time-pressure, behavioral nudges, and the architecture of situations, all play a role in shaping judgment and bringing down the premise of a rational unconstrained self. While Mill would respond that it is precisely for these reasons we should nurture alertness and generate unexpected alternatives, this does not marry well with experiences of everyday life that do not always lend to deliberation.

As a systemic norm privacy mediates connections, but not necessarily for the purpose of being alone or secluded. Indeed privacy may involve being more open, but with a selected set of people or organizations. The principle then is one of management and modulation. For example the gay community has historically been forced to operate away from the public because of societal antagonism and the legal status of homosexuality (Schoeman, 2008 [1992]). This leads us to the observation that while autonomy is good (independence from social pressure) by framing privacy in terms of seclusion and being alone, we are also in the terrain of isolation.

Instead a more dynamic view including contexts, choice and modulation of connection helps shift the focus from 'the right to be left alone' to privacy as that mediating principle that facilitates more enriching relationships. Rather than seeking isolation and seclusion, we might better manage our connections and move more freely by dint of our associations. Connecting with ideas about management and access, this sees control arise because of what groups and selected associations afford—the pursuit of valued ends without unreasonable social sacrifice.

Transparency and the economics of privacy

Transparency has a number of roots, sometimes with contradictory implications and consequences. For example liberal democratic principles are based on transparency, and as the foremost contemporary liberal philosopher Rawls puts it: 'Principles are to be rejected that might work quite well provided they are not publically acknowledged, or provided the general facts upon which they are founded are not commonly known or believed' (2005 [1993]: 69). While clearly there are situations where the state acts on our behalf and without our knowing the details, the principles that inform these actions should be transparent and public. Bobbio (1989) speaks directly to the breakdown of transparency in democracy characterizing the [Machiavellian] power of the prince of today as more effective when it is hidden from citizens. There are functional dimensions to this as power is more readily, rapidly and successfully exercised when power is unforeseeable and capable of surprise. A non-democratic approach to transparency also means that the prince does not have to engage and domesticate a public too easily swayed by passions and who are unable to form rational opinions. For Bobbio we have arrived at a situation characterized by absence of transparency, the uncheckability of power, closedness to the public, non-publicity of decisions and the creation of 'a principle of state action which breaks the moral law against lying' (1989: 19).

Utilitarianism itself has an interest in transparency that sits outside the democratic interests of liberalism. We might draw attention to Bentham's interest in transparency and extension of the visual field. In such vision:

> A whole kingdom, the whole globe itself, will become a gymnasium, in which every man exercises himself before the eyes of every other man. Every gesture, every turn of limb or feature, in those whose motions have a visible impact on the general happiness, will be noticed and marked down. (Bentham, 1834: 101)

The Foucaultian (1977) connection with Bentham, panopticism, affect and visibility is well-understood, but we might pay closer attention to the latter part of the second sentence in the quote and the nature of impact on the 'general happiness.' Transparency is intended as a positive concept to promote societal net benefit and general happiness, not as a punitive measure. Transparency and surveillance (or more accurately, equiveillance) in this context is positive and accords with the Enlightenment doctrine of making all things present so to generate understanding, and make life better. Bentham's intention is not sinister but aimed at a net improvement (in terms of happiness and pleasure) for all. On citizen transparency Posner (1983) makes the somewhat controversial, but admittedly refreshingly clear, argument that people's privacy protection is economically inefficient. Being controversial he is refutable, but usefully his is a contribution and field-mark to help map and navigate writings about privacy. His Benthamite views are built on the idea that personal information is less valuable than many liberal-inspired commentators suggest. He goes further suggesting that breaking down privacy domains and promoting transparency may be beneficial both economically and morally. His overall approach is to employ arguments derived from economics and its interest in wealth maximization as a means of assessing the value of privacy. Bringing together economics and justice, Posner defines the latter as 'the way in which the major social institutions distribute fundamental rights and duties and determine the division of advantages from social cooperation' (1983: vii). This sees common law as best explained and handled as if judges were trying to maximize economic welfare. Posner's argument is predicated on rational maximization and is not confined to regular market transactions as the conceptual apparatus also explains, for Posner, non-market behavior. On opening up and making transparent to the market, there is an argument to be made on net utility. This is best approached in smaller scale terms. For example, we might ask why one would want rights against a society that exists to enhance, preserve, defend and promote co-operative life? Somewhat ironically, as accounted for in Chapter 11, this is the same argument that Marxists make about privacy and living in communities with others.

Moral truffles

There are then at least two forms of transparency: 1) *liberal transparency* and the opening up of machinations of power for public inspection; and 2) what I term here as *radical transparency* that opens up both public processes and the private lives of citizens. Sharing commonalities with the 'artificial barriers' argument I disputed in the *Mood of Information* (McStay, 2011), Posner (1983) suggests that privacy should only be protected when access to the information would reduce its value. This reflects Cohen's observation that 'within a liberal market economy, it is an article of faith that both firms and individuals should be able to seek and use information that (they believe) will make them economically better off' (2012: 9). This involves a conflation of Enlightenment ideas about morality and democratic transparency, reason and academic openness, and the shift to free markets.

Preferring corporate and organizational privacy to individuals' privacy, Posner argues that allowing individuals privacy is not defensible as people use privacy as a means to conceal or selectively disclose information. Thus where privacy is often articulated as a means of control over self-presentation, and the management and sharing of information, these are grounds for Posner's criticism of privacy. Where a more positive account of autonomy involves control about how to present and stage ourselves, to whom and in which contexts we want to do this, this is the very focus of Posner's ire. For Posner privacy is very much connected to the negative ideas of withholding and concealing, particularly in regard to personal uses of information. In building an account of transparency he points out that seeking to control the flow of information is a wish to control others' perceptions. Comparing this to marketing he argues that if it is wrong to withhold information or misrepresent goods why it is acceptable for people to do the same? To quote: 'It is no answer that people have "the right to be let alone," for few people want to be let alone. Rather, they want to manipulate the world around them by selective disclosure of facts about themselves' (1983: 234). Conversely, Posner argues that businesses should be afforded greater levels of privacy so to encourage innovation and because placing businesses under the public spotlight harms growth. Where most will elevate personal privacy and downgrade the need for organizational privacy, for Posner this is strange economic behavior. This leads him to contend that 'the case for protecting business privacy is stronger than that for individual privacy' (1983: 249). Personal privacy in this account leads to distortion, misleading and manipulation, and therefore does not maximize wealth (and thereby the net value of the economy and by extension utilitarian happiness in society). While many of Posner's ideas about privacy favor data managers, corporations and sometimes governments, the economics of privacy argument becomes less easy to criticize when we make the experience of this more local. For example, if a healthy person seeks to obtain life insurance then this person may want to have all aspects of their body and health scrutinized. That is to say, we may seek differential pricing based on high levels of personal information when it works in our favor. Rosenberg (2000) makes a similar point in his criticism of rights. Arguing that privacy is relativist and lacking commonality among the cases in which it is cited, he argues that it proves an annoyance to merchants wishing to process information about us. Likening privacy to individual taste, he remarks that making privacy a moral foundation for social institutions 'will be no more fruitful than seeking a moral foundation for truffles' (2000: 76).

Teleology

Posner draws on Bentham's work on utility (who drew directly and indirectly from Cumberland, Hutcheson, Priestley, Helvetius and Beccaria). Posner highlights there are flaws in Bentham's argument, not least that the happiness of millions cannot be measured and competing policies to reach pleasure maximums cannot be aggregated and judged. Less relevant, he also takes Bentham to task for his views and assaults on language (Bentham had a very conservative view of language and was not keen on simile, metaphor or embellishments and sought to expunge these). Posner also criticizes his ideas about prison reform, brainwashing, compulsory self-incrimination and torture, among other themes connecting to totalitarian regimes. However, having worked through his criticisms, what he does take from Bentham is his interest in rational maximization and the meta-lesson about transparency. Drawing on Bentham's *An Introduction to the Principles of Morals and Legislation*, Posner highlights Bentham's two poles of governance and orientation (pain and pleasure), and that: 'Men calculate, some with less exactness, indeed, some with more: but all men calculate. I would not say, that even a madman does not calculate' (Bentham, 2000 [1781]: 146).

Posner and Bentham's arguments are at heart teleological and consequentialist accounts where ends or consequences determine whether an act is good or bad (thus departing from many liberal deontological principles in which acts are inherently good or bad). For total market efficiency, transparency and net societal gain to be reached, goal-directed systems should be allowed to work optimally and without disruptive intervention. This teleological position (my phrasing, not Posner's) extends as far as considering the long-term value in full disclosure of sexuality, political affiliations, minor mental illnesses, early dealings with the law, credit scores, marital discord and nose-picking (all Posner's examples). These are all situations that might create irrational reactions from prospective partners or employers that might be considered for withholding from public view. Posner sees differently making a virtue out of disclosure and that if radical transparency were the norm, over time irrational shunning and biases would be excluded. Market systems, for Posner, are that which weed out irrational prejudices. While those with an interest in protecting privacy tend to argue that private communication is required to ensure the free exchange of ideas without embarrassment, fear of becoming public property and that society is richer with individuals who have occasion to develop themselves privately, Posner sees otherwise instead promoting what is dubbed here as *radical transparency*. This will-to-openness operating under utilitarian auspices presents us with an ideal where there is no need for private domains (with privacy for businesses excluded). Such a scenario clearly has totalitarian character and while privacy *does* allow people to hide misdeeds and abuses, this is far more preferable to radical transparency, or forced transparency, and one-dimensional moralizing. This would provide social risks for many groups and to offer only one criticism of radical transparency, it calls for a significant sacrifice on behalf of some to enlighten the rest of us. It is not clear what net positive results would be gained by such a request or regime.

Conclusion

This chapter has accounted for utilitarianism and privacy in two key ways. It has highlighted Mill's *On Liberty* as a philosophical manifesto for privacy for those who see privacy in terms of seclusion. Criticisms were made of this, particularly in regard to over-emphasis on individuals, being alone and what I argue to be the *seclusion trap*. A more positive account of privacy is preferred in this book that reflects the means by which we manage and modulate connections with others (which at times involves greater openness). The second strand to this chapter runs from Cumberland through Bentham and on to Posner. This involves the relationship between calculative rationality, utility, transparency and economic judgment about the value of privacy. While transparency as a principle seems to be indexical with a good thing, the consequences and externalities of *radical transparency*, or forced transparency, are deeply undesirable because resistance becomes tantamount to guilt, and the putting into practice of transparency would require use of power along with the stripping of choice and autonomy. On the question of whether transparency might ever represent a net gain, the answer is no as it would simply promote greater self-censorship and less openness. Indeed the will to public transparency, the removal of artificial barriers and the positing of privacy as a moral truffle pervades industry accounts of privacy—as witnessed in the intense lobbying for more relaxed privacy laws in Europe broached in the last chapter. The need to make 'present-at-hand' and the wish for all to be visible and subject to a common language of exchange is very reminiscent of Heidegger's concern about a metaphysics of presence (that is addressed in Chapter 6). Questions might also be raised about the nature of transparency. Transparency is less a thing but rather has more in common with cleanliness—a virtue (Birchall, 2012). While connected in part to democratic functioning its impetus is farther-reaching because it is concerned with maximizing reason, clarity, visibility and ridding society of superstition (Kant, Rousseau and Bentham all figure within this). The interest and will-to-transparency is itself worthy of remark and as Bennington (2011) in discussion of Kant's (1983 [1784]) *Perpetual Peace* suggests, transparency itself acts as a kind of veil, or what we might loosely infer as an ideology. The question is less about a continuum or dualism of transparency and privacy, but rather what is it that transparency upholds? Phrased otherwise, it is akin to be being blinded by light so for shadier procedures to take place. For Bennington transparency acts as a kind of veil, or fold of secrecy. To give an example the UK's Prime Minister, David Cameron (2011), penned an article for *The Telegraph* in 2011 aptly titled *We are creating a new era of transparency* remarking on the need for increased information on government spending as it lets 'people hold the powerful to account, giving them the tools they need to take on politicians and bureaucrats' and helps citizens make 'informed judgments about their future.' While these are good practices and fine principles, the revelations about the UK's surveillance program that emerged in 2013 by means of the Edward Snowden leaks highlights the fact that transparency is used as a tool for strategic political communications and as a veil. The net effect is to be misdirected by openness.

CHAPTER FIVE

Pragmatism: Jettisoning normativity

Chapters 2, 3 and 4 respectively assessed border-based, liberal and utilitarian approaches to privacy. This chapter considers privacy in relation to pragmatism, that creative area of philosophy that revels in indeterminism, prefers to look forward, and refuses to be restricted by dogma, strictures and the preeminence of monological reason. Pragmatism is a refreshingly clear approach with simple premises, yet high levels of critical engagement with sibling philosophies. Indeed a common characteristic of much pragmatist writing is its accessibility. It shares sympathies with liberalism and Mill, non-reductionist strains of utilitarianism that reject fixed norms, and that general tendency to ask questions about what will bring about a better world.

In this chapter I pay particular attention to Rorty's (1989) *Contingency, Irony and Solidarity*, while acknowledging the tradition of pragmatism through terms and writings provided by William James and Dewey. Rorty's interest is the relationship between metaphysics, rights and morality, and the ways in which we engage with these today. The background to this concern involves contemporary emphasis on textualism and that mode of social criticism that requires we understand the root yet contingent discourses of our beliefs. This raises a number of questions: is an absence of metaphysics politically dangerous? Centrally, should we allow rights, norms and in our case privacy to be contingent upon context? Do we a need a deontic conception of privacy to prop it up, or can we open it up for contextual redescription? To generalize, pragmatism eschews normative approaches in favor of situation and ethnocentric answers to local problems. It is almost a *volte-face* of Kant, the hegemony of reason, and his argument that local and relative interests should not determine norms. In regard to privacy the proposition that local conditions, wishes and specific actors might determine privacy norms is encapsulated in Nissembaum's (2004, 2010) contextual approach to media. This eschews binary terms (access/denial of access, open/shut, public/private and so on) preferring to express privacy in term of appropriateness, context, the type and nature of information, or what and with whom information is being shared. This chapter thus uses pragmatism as a means of unpacking and exploring Nissenbaum's approach, and as a way to account for its implications, particularly in regard to deontology and questions on consent.

Being contextual

Nissenbaum's notion of contextual integrity dissolves absolute privacy norms through the observation that people do not require complete privacy and different norms apply in different circumstances. This is a reaction and attempt to deal with the fact that privacy matters are informationally and technologically complex. Contextual integrity is also a recognition that matters have always been contingent upon who is involved, what the information is, how it is transmitted, who might be affected if a breach occurs, and the difficult ways in which social contexts can overlap (e.g. should one tell one's boss that his/her spouse is having an affair [Nissenbaum's example]). Her argument involves recognition that different social situations involve different self-generated norms, with these often being quite unique and different from each other. Such observations allows us to recognize that privacy norms are not once-and-for-all objective, but neither are they individual. Instead they are culturally (contextually) specified by means of realities co-constituted by those living in them.

Her concern is with the appropriateness of information flow that is established in reference to a range of factors including roles, activities, norms and values. Respectively, these involve: *roles* and the capacity by which people act in context, e.g. student, lecturer, medical receptionist, waiter and so on; *activities* in terms of what people are seeking to do or achieve, e.g. buy a book online, vote, report a lost wallet or share a photo online; *norms* that define the duties, expectations, actions, privileges and what is acceptable for a person or technology fulfilling a role; and *values* that refer to the goal of a system or social arrangement. This recognizes that we are always trying to achieve something, whether this be obtaining education, staying healthy, being social, or participating in democracy by voting. The test for any new privacy or informational practice to be introduced is to measure it against existing practices and check whether it tallies with the objective of the system. This means assessing the extent to which it violates or complies with the existing norm. Here Nissenbaum shows her pragmatic sympathies, remarking that we should assess 'each new practice on its individual merits [and] in comparing it with entrenched practice it merely requires better performance on a cost-benefit analysis' (2010: 179). This tallies with James' (2000 [1907]: 28) often quoted, yet mostly distorted, observation on the cash value of truth. Less a point about utility or putting to work of truth, James' message is that truth is an evolving notion, that nothing is necessarily true and truth is subject to change. Although there is a conservative tendency in contextual integrity that favors the existing arrangement (a point Nissenbaum readily acknowledges), the challenger should be accepted if the new practice is an improvement and more effective in achieving the aims, purposes and values of the system. Indeed, while I connect Nissenbaum with pragmatism in this chapter, there are strong links also to be made with strains of rationalism and utilitarianism given her interest in measuring moral propositions. Her objective then is to create means by which new informational practices can be assessed in relation to existing ones. In a manner reminiscent of cybernetics, this is done by means of comparing the proposal with the objectives and values of the context (e.g. better physical health or raising educational attainment) and subsequently making changes if the proposal advances the context's objectives. Assessment is carried out not by imposition of universal privacy norms, but by recourse to 'the backdrop of the specific ends, goals, purposes, and values' (2010: 171). Privacy norms thus are driven not by external normativity, but by local values and objectives of contexts.

For Nissembaum, and the systematic account being developed in this book, a one-dimensional approach to privacy does not cover our own multiple privacy scenarios, never-mind those of others. Addressing the need for greater flexibility, Nissenbaum introduces a more neutral account of privacy, or the recognition that we can talk about privacy and media without having to make recourse to normative moral terms that tend to be absolutist in coverage. Nissenbaum, following Gavison (1984 [1980]) and her work on neutral privacy, highlights that approaches seeking to better understand context need not deny normative judgments, but that neutral and systemic approaches better characterize how privacy works. This leaves us free to describe and account for privacy events without having to make moral judgments each time we encounter a situation. While keen to play down absolutism, Nissenbaum and Gavison's approaches to privacy are oriented by liberal ideas of privacy that involve autonomy, and freedom for self-development and experimentation without external pressure or censure. Following Gavison too, we might better see privacy in sociological and networked terms (although Gavison's discussion is about law), and that which uses the language of access, control, proximity and local privacy conditions. Much of this is reminiscent of Altman (1975), and Altman and Chemers (1980), who wrote extensively about the modulation and regulation of dialectical conceptions of privacy, along with accounts of control and management of privacy norms in changing circumstances characterized by different actors and environments. Providing an antecedent for contextual social thought, for Simmel (1906) norms emerge out of settings, although in each situation we develop 're-serve,' or that capacity and need for withdrawal. In general this involves a dynamic and ongoing negotiation of borders and maintenance, subject to the development of circumstances, needs and wishes. There are important consequences of this contextual approach, not least that this underlines one of my own arguments in that privacy is not just about seclusion. This is because as well as requiring respite from others, privacy also involves making ourselves more open to chosen others. Privacy in this more flexible approach involves practices of optimization and negotiation of access that, to reiterate, involves seeking interaction as well as restricting it. This positive view is reinforced through the principle of choice and autonomy. This has a performative dimension reminiscent of Goffman's (1990 [1959]) famous discussion of the presentation of the self in everyday life. Playing between contexts, highly competent performers are aware of the codes and conventions of contextual use, and are alert to goings-on and shifting interaction among co-performers.[1] Likewise, the skillful performer possesses a range of faces, identities and management techniques with which to carve up the world in terms of behavioral and interactional modalities, or ways of being (reminiscent of management and privacy as discussed by Petronio, 2002). Contextual integrity is both creative and pragmatist in its critique of deontological accounts of privacy because it recognizes situation dependency, and that privacy norms emerge from interaction (and are not just about seclusion).

To develop this approach is to consider privacy in a different guise from what we are used to. Privacy is generally thought to involve withholding information or something of the self; a personal drama in which some long-held secret is disclosed; and even a macro-breach where a country is found to be colluding with the world's leading social networks for the purpose of surveillance of non-national citizens (as highlighted in the Edward Snowden leaks of 2013). In a sense, a contextual and systemic approach to privacy is much more boring as it is about the status quo, equilibrium and the optimization of inter-relationships between people and other entities within a given system. Privacy is not just about when something goes awry, but involves the principles and norms that dictate optimal arrangements. Further, in a pragmatic and contextual reading, these norms emerge from the entities within the arrangement. A privacy event then can be said to occur when protocol, or a negotiated or implicit set of agreements is breached, although this need not involve us knowing. As with Rorty's approach, this is deeply local and while Nissenbaum does discuss normative approaches, this is not her focus. Analytically her system involves understanding roles, activities, norms and values of a system, and this provides a flexibility not found in the absolutes of deontic, paternal, utilitarian or dualistic conceptions of privacy. This progresses us to seeing privacy in terms of protocol and emergent norms that guide the functioning of informational arrangements. By shifting towards protocol we can easily extend emergent norms to non-informational situations involving privacy. For example, university students on moving into a shared house both contribute to the privacy norms in operation within that household, yet are also determined by them. The point is that the norms are not imposed, but they are either silently or vocally, or tacitly or explicitly, negotiated.

The problem with normative ethics

Norms are not just beliefs, but also forms of behavior that we think others should engage in. As a set of standards norms are public rather than private affairs. A normative account of ethics involves an attempt to state principles about what is considered good and best. It is an attempt to define what is right and wrong, and by extension the nature of duties and what should be forbidden. Often decisions about what is both good and best are not straightforward. Kegan (1992) uses the example of the decision to save someone drowning. While clear that this is the 'good' thing to do, what happens if a boat has to be stolen to do this? This involves a decision—should one save the life, or contravene property rights? Few would walk on comfortable in the decision not to appropriate the boat, but the example serves the function of highlighting that moral decisions are not always clear and conflicting moral dimensions come into play during a decision. The weighing-up of all of this involves agreement on what dimensions are real and relevant, understanding the implications and consequences of these dimensions, and how they interact and conflict. Sometimes these decisions are not easy to make and norms may be balanced. The case of the public's right to know versus privacy is a good example of this, as mentioned in Chapter 1. Absolutes and principles are easily defended in theory, but day-to-day circumstances have a habit of providing situations that upset clear-cut philosophical axioms. To offer a privacy example, a government initiative in India illustrates well the problem of deontological approaches to privacy. Possessing little in the way of privacy legislation (compared to Europe), India has rolled out a national biometric ID scheme. This has helped those in poverty gain a recognizable identity in lieu of passports, driving licenses and proof of address, and allowed citizens to receive benefits, and possess medical and school records. The example highlights positive benefits to information gathering, yet it also involves massive collection of data highly useful to Indian state capitalist businesses, and it is reasonable to ask whether the interests of the state as protector of citizens and state-owned businesses can be kept separate.

Pragmatism is not the first area of philosophy to question normativity. Indeed, Spinoza saw a normative vocabulary as at best a useful fiction, and more negatively as a source of confusion that only exists because people are too limited in their capacity for knowledge and understanding (Schneewind, 1998). Beyond criticisms of fiction, normativity is difficult to live with. As numerous heads of technology companies like to remind us, privacy is being subject to renegotiation in novel socio-technical contexts and is no longer a 'social norm.' This is a silly opinion because privacy is a far too basic principle of behavior for it somehow to no longer apply. However, the fact that protocol is subject to flux has to be taken into consideration. Clearly aspects of our environment and the ways by which we connect with others are changing, and this requires we reconsider norms. To suggest however that privacy is not important is to misunderstand the reach of privacy and the ways in which privacy norms are fundamental to how we relate to each other—whether this is mediated through a social network, gendered door entrances, or simply by withholding information in a face-to-face interaction. The proposition that privacy is unimportant can be readily rejected when we recognize how far privacy pervades, constructs and mediates our social lives. Privacy then is akin to language in that its norms may change (and communications companies and other stakeholders do have a significant role to play in establishing protocol), but privacy is highly unlikely to disappear as this would leave behind a very strange and undesirable world. Instead we need an approach that recognizes the value of privacy but allows flexibility and change—as with language. A pragmatic worldview is useful in this regard as antipathy to the status quo and normative ethics is deeply reflected in pragmatism. As James (2000 [1907]), Rorty (1989) and others remind us, the pragmatic view is free of absolutes instead seeing both moral and epistemological truths as human and contingent upon period and context. In this view philosophy is humanity's 'ongoing conversation about what to do with itself' (Rorty, 2007: ix). Beneath this simple premise is a more significant assertion that highlights the compliancy of reality where 'Man *engenders* truth upon it' (James, 2000 [1907]: 112 [emphasis in original]). Our truths are not found but they are created. Dewey (1995 [1908]) similarly observes that whereas traditionally knowledge is taken to be a reflection of the world, instead the subject plays a much greater role in constructing the world. That is to say, our knowing directly participates in the formation of the world giving reality a 'practical character.'

Changing norms

The relationship between pragmatism and privacy is a difficult one. Refusing absolutes, Rorty (2010 [1993]) deconstructs the notion of human rights as a deontological premise, arguing that the notion of foundationalist human rights is outdated (involving self-justifying premises that offer foundations to later premises). Rorty's pragmatism requires that we do not go searching for an external reality of rights, but get on with addressing the constructedness of norms, principles and their appropriateness to the situations in hand. Expressed rights written into policy are for Rorty just that—expressions of a broadly agreed viewpoint. They are a set of demands and norms we make on and for ourselves within a given set of social conditions. They are not a description of what *is* right (this would be deontic) but are an ethnocentric expression of what is deemed right given all the circumstances taken into consideration. To make this point Rorty (1998: 171) uses the example of the US Supreme Court's construction of a constitutional right to privacy that he sees as an expression of a self-conscious decision about the right thing to do, rather than as a demonstration of transcultural superiority. The refusal of foundationalism is a key argument that for Rorty is established by moral intuition, which stands in contrast to assertions of moral knowledge, or claims that pertain to be able to correct the intuitions of a community. The reason why there cannot be moral knowledge is because as a proposition it is founded on a correspondence theory of truth. Norms for Rorty are agreed, not found. Specifically addressing the topic of human rights, Rorty's position is that recognition of such rights conforms to the way well-off people from the First World should treat each other, rather than reflecting some basic reality. This means they are not transcultural but ethnocentric and this obliges us to reject the premise of a view from nowhere. However, to point to rights as relative or constructed is not to deny their power of affect, or importance to us. Neatly phrased, Rorty (2007: 196) comments that 'a norm is just a kind of fact—a fact about what people do—seen from the inside.' This sense of self-determination connects with liberal and utilitarian ideas, not least Bentham's depiction of morality as that which is determined by the people it is to govern. This also connects with Kant who as detailed in Chapter 3 returned moral being from 'out there' to that which is within us. While Kant (1991 [1785]) does not go as far as framing this in terms of historical contingency, he fulfills the larger task of connecting morality, autonomy and choice. Rorty (1998: 172) directly affirms this viewpoint (and presumably knowingly leaving himself open to deontic charges) by arguing that morality is a question of efficiency as it is about finding the best means to 'bring about the utopia sketched by the Enlightenment.'

Rorty's (1989) ideal in *Contingency, Irony, Solidarity* is the liberal ironist who faces up to foundationlessness and the premise that there are no metaphysical foundations for the liberal concern with justice. This is to recognize the ethnocentric nature of liberalism while seeing that there is no God's-eye or Archimedean point of view to which to take recourse. Indeed, Rorty's self-stated aim is to convert ironists into liberals and liberals into ironists. To do this he makes a distinction between being public and private, pointing out their irreconcilability. He rejects both the Platonic sense of how self and state might be fused (Plato's *Republic* is dedicated to promoting harmony of the state and the individual, seeing division fostered by the conflict of private interests with those of the state as the ruin of this relationship), and the Christian claim based on perfect self-realization so to serve others. Instead Rorty splits philosophy with what he characterizes as private philosophers on one side (e.g. Foucault, Nietzsche, Kierkegaard and Heidegger among others), and those of a more public form on the other (e.g. Marx, Mill, Dewey and Habermas) with a greater interest in the public good. The former private approaches are those that deeply privilege autonomy, self-creation and respite from external events; the latter are involved in social efforts to conceive other worlds. While in privacy circles Foucault (1977) is recognized as providing key explanatory mechanisms for surveillance practices (in regard to the Benthamite means by which the few may watch, influence and control the many) by means of analysis of dispositifs (structures that help maintain power over bodies), he is also a privacy philosopher because of his rejection of the public 'we' and his interest in subjectivity. The autonomy that Foucault seeks is not that which could be scaled up to policy. Indeed, Rorty (1989: 83) accuses Foucault of being 'pretty much useless when it comes to politics,' but Rorty also takes Habermas to task for being overly critical of Foucault. Where Habermas (1990 [1985]) sees Foucault (along with Hegel and Derrida) as overly subjectivist and antithetical to progressive social hope, and philosophy as means of providing a public/private social bond (by means of rationality), Rorty remarks that Foucault's project of the self is a deeply necessary and valuable one as it highlight effects of institutions on people. Rorty (1998) points out that where Habermas sees Foucault as a bad public philosopher, he sees him as a good private one. For examples we can point to Foucault's (1990 [1976]) work on sexuality, historicism, genealogy, the construction of self-scrutinizing subject, and the aforementioned surveillance work and experiences of self in a surveilled state. Indeed, it is in the politicization of intimate domains of everyday life that Foucault does privacy studies the best service. This is achieved by means of innovative methods to assess the construction of the self so to be able to put forward new forms of self. We might also see a real link here between Foucault and Mill (which Foucault would reject), with the latter also being interested in the self as an ongoing private concern separate from both state machinery and less overt social influences. Such ironism (recognition of contingency) and lines of questioning must be maintained.

Rorty (1989: 75) identifies phrases routinely used by the ironist to include *Weltanschauung*, perspective, dialectic, conceptual framework, historical epoch, language game, redescription, vocabulary and irony; but for the purpose of this chapter we should also add the word 'context.' Rorty's answer to the incommensurability of private and public philosophical projects is a simple one: ensure the former stays privatized (and beyond the reach of the public) so to be able to maintain more general liberal values (e.g. avoidance of cruelty). This is not as improbable as it may sound as we all mix with different groups and communities, with different norms. A single person can have more than one outlook. The problem is one of emancipation, as those with a primary interest in the self do not believe in the possibility of collective emancipation; and those interested in public structures do not see the infirmness of their metaphysical principles. Foucault's criticisms of public philosophies and the possibility of autonomy are not without flaws. As Rorty (1989: 63) points out, 'he is not prepared to admit that the selves shaped by modern liberal societies are better than the selves earlier societies created.' He continues that while Foucault points out processes of acculturation and constraints placed on modern life (not least his work on discipline and apparatus of social power) he does not answer the question of whether this is compensated by a decrease in pain. Rorty does not suggest that all is well, but rather sees liberalism as an ongoing project continually remaking itself and that it 'already contains the institutions for its own improvement' (ibid). For Rorty the achievement of private understanding and public citizenry cannot be fully combined into a single vision. Nor should one be privileged over the other, but rather they are different irreconcilable categories of philosophy. While we might possibly have a knee-jerk reaction to this seeing the public as good and the other as somehow indulgent, the philosophy of the private and what Rorty refers to as the 'ironist' position sees the contingency of metaphysical beliefs (by means of historicism and ethnocentricism). This is a necessary critical activity and the ironist is the person who is able to face up to the contingency of their own wishes and beliefs, rather than take recourse to some foundationalist metaphysical once-and-for-all-and-everywhere principle.

It is less about indifference to pain in the world than a critique of the proposition that there are natural answers, or algorithms guiding and ranking for us how we should live. If we remove the metaphorical and metaphysical bicycle stabilizers, then we are left to orient, balance and progress ourselves. Rather than the truth of a matter being subject to correspondence, truth instead comes to be by means of open debate and discussion so for us to decide the best way of dealing with current contexts. This should not be difficult to digest because, after all, societies tend to bind together by interests, hopes, fears and language rather than metaphysical beliefs. This means that norms necessarily become more contextual and creative in that sense of recognizing indeterminacy, living with it, and making the best of it. It is a rejection of reason-based Enlightenment metaphysics and the proposition that there are *singular* moral truths (classically, reason can only take us to individual answers).

Rorty (2007: 29) elsewhere tells a similar story by taking recourse to Nietzsche's *The Gay Science* and discussion therein of the pre-Socratic Greeks who were allowed a plurality of norms because belief in one god did not equate to a denial of another god, nor blasphemy against either god. This is a key point for developing the general character of contextual approaches to privacy as belief in one norm in a given situation is not a denial of another set of norms. This eventually (if not already) leads to charges of relativism and Rorty's (1989) answer is to change the terms of the discussion (redescription). What is relative for Rorty is not tantamount to being valueless, meaningless or worthless. This is because rights do not need to be externally validated as correct, but instead we understand them to be local and produced by a configuration of circumstances. The liberal ironist thus recognizes the local, contextual, contingent and, yes, relative position of his or her argument, yet stands-up for them anyhow. A belief can be fought for and considered as deeply important, even when we recognize that it is contingent and contextual. This is not a denial of principles, but rather of metaphysical props. It is a rejection of the scientism associated with the Enlightenment and once-for-and-all answers towards continual redescription. The somewhat utilitarian way to assess this is whether what is proposed is better than what was. Where relativism may smack of amorality, the argument instead is to see morals in terms of community values and practice, rather than as externally generated norms.

On privacy itself, Rorty questions the extent to which what is private can and should be linked with what is public, positing that 'there is no way to bring self-creation together with justice at the level of theory' (1989: xiv). There can be no unifying theory or theoretical discipline that can encompass both the public and private philosophies, but 'The closest we will come to joining these two quests is to see the aim of a just and a free society as letting its citizens be as privatistic, "irrationalist," and aestheticist as they please so long as they do in their own time—causing no harm to others and using no resources needed by those less advantaged' (ibid). A key question then is whether we should take Rorty at his word on contingency, and allow privacy to be determined by the actors involved. If contingency grants greater rights for communities to decide where-next for themselves, and for norms to be self-generated rather than drafted in from history 'out there,' can such self-guidance handle more aggressive actors with more resources at their disposal? If contexts can be changed by intensive lobbying, barrages of rhetoric, repetition of messages and snappy phrases that tap into zeitgeists and hopes/fears ('privacy is no longer a social norm'), corporate educational programs (for example on cookies and online advertising that tell only one account of the story) and other means of persuasion, should we be somewhat more concerned about how contexts come to be? Related, might truly contingent and contextual approaches jettison our privacy props too readily? Do deontic and long-standing principles afford protection from trends and reactionary moralizing? Further, in regulatory and practical terms, with what regularity should we be able to recreate? Finally, how comfortable are we with the premise of loosening the deontic reigns on these norms that so deeply affect European thinking and the European Commission's legislation on privacy. If we move towards a more contextual and local approach, who and what might we be letting ourselves in for?

Conclusion

As depicted, a contextual approach shares real similarities with a Rortian account of privacy, particularly in regard to contingency, rejection of fixed/deontic/paternal norms, looking forward rather than backward, and that local sense of making and building. Indeed, this is very much about self-determinacy. However, does Rorty's account play-out as a public/private philosophy? We might use Rorty's own approach to consider this. Rorty splits philosophy down the line of those involved with projects of the self (e.g. Foucault) and those interested in grander societal visions (e.g. Habermas), but what does a contextual and pragmatic approach to privacy look like if scaled up to public-policy level? This is a real question, particularly given Nissenbaum's interest in media that occupies center-stage of European wrangling over principles of data processing. What then would these approaches deliver in practice? While I have sympathies for these approaches in private, I wonder if in the public rejection of universalism we jettison the deontic dimension of privacy too readily leaving it prey to those able to deeply influence contexts. Put otherwise, if we were to adopt in full a contextual and pragmatic account of privacy norms we invoke bi- or multilateral agreements, reciprocal relations and local [private] regulation. However, in doing so, we invoke a situation where we weaken the possibility of overt sanction, effective power and a sovereign force able to engage with corporate actors. Perhaps a little guiding public normativity isn't a bad thing?

SECTION TWO

KNOWING

CHAPTER SIX

Heidegger (Part 1): Concerning a-historical being and events

In the preceding chapters I have addressed some of the ethical dimensions of privacy in relation to philosophy. This was done by means of considering the arrangements by which we live together, the formal character of privacy, and the consequences of moral and political systems for privacy theorizing; and by promoting a broadly liberal and pragmatic view, albeit with warnings and caveats. We now proceed into the second section of the book cautiously open to dynamic privacy protocol, and driven by a strong sense of the need for true consent in a mediated setting and that privacy is neither an artificial barrier nor a moral truffle. Instead by means of the broad background depicted, we see that privacy lies at the heart of the most basic principles of interaction and it is in this observation that a dynamic, systemic and protocol-based approach finds its origins.

If privacy is about managing insights, information, access and perceptions between a multitude of human and non-human actors (e.g. technical systems), what can philosophy tell us about the state and conditions of knowledge in these arrangements? Moreover, what of the understanding, being and experience of privacy itself? These are all points of investigation for the next seven chapters and we begin these by considering Heideggerian philosophy in this chapter and the next. Extended treatment of Heidegger is warranted because of the diversity of his writing and the multiple ways in which he is relevant to privacy and media matters. This chapter accounts for his ideas about technicality, metaphysics and being. It considers the implications of calculative rationality, making present, excavational technology and what I designate as *a-historical data mining*, or that realization and formalization of absolute visibility and radical transparency introduced in Chapter 4. The chapter closes by assessing Heidegger's approach to events. Phrased in a deceptively simple manner, an event involves consideration of how things come to be for people, and while I am sure real Heideggerians will take issue with this gross generalization, it serves for our purposes. The intention of this discussion of events is to invite consideration of the extent to which heterogeneous behavioral technologies are able to engage with the 'whatness' and 'suchness' of human experience.

A philosopher of seclusion

Heidegger is an introspective philosopher and radical phenomenologist who believed that the private realm of the individual is the means to truth and that great courage is required for solitude.[1] He might then be said to be a philosopher of seclusion, if not privacy, not least because of his famous vacation at home on the edges of the Black Forest in Germany where much of his thinking was done. Although preferring solitude, he has no interest in liberalist philosophy. In an invective against liberal norms Heidegger speaks of the liberal subject in terms 'of the emergence of the most insidious semblance of autonomy' who 'speaks only of "freedom" and liberation.' Real scorn however is reserved for liberalism's self-absorption and how, by its establishing of deontic norms, declares humanity as the centre and mission of history. Instead for Heidegger: 'Historical making inceptually knows no mission, since it has no need of one, having been consigned enough in the arrogation of the truth of being' (2013 1941–2 §184: 133). What some see as a liberal enlightened ascent, he sees in terms of egoism, fall and loss, because by adopting the mission as ordained we promote a self-absorbed worldview with modern tendencies at its center.

Heidegger is the foremost philosopher associated with critical thinking on technology. Among these reasons, he stands out because he refused the common sense notion that a technology is simply a tool to achieve a task of some sort. Whether these involve global communicational networks, plungers to unblock drains or triangular chunks of flint, the principle for Heidegger is the same in that technology is best seen as determined by contexts that give rise to the technicity of an object in a particular arrangement. For example, on my desk a heavy oblong piece of slate (2cm × 2cm × 20cm) acts as an excellent paperweight to sit astride open books I am typing notes from. The technicity is not embedded in the object, but it comes to be through context and an object's place within a given reality. To understand the being of things is thus to inquire about the milieu in which it is identified. This is a hermeneutic act in that the being of an entity is to be understood in the context of which it is disclosed. It is less about *what* than *how* things come to be. For Heidegger then there is a duality of thing and context, and it is the latter that is of far greater importance for a philosophical approach to technology and privacy. Possessing separateness, technicity also involves a unique sense of rationality and preeminence over human life. For Heidegger however the intensification of the technical environment in modernity involves a shift from acts of fashioning and making (generally with other tools) to larger technical enterprises that have, to an extent, come to hold us captive. While a surface consideration of these ideas may involve a rejection of Heidegger out of hand as a technological determinist (a very unfashionable view to hold), if we consider we are born into technical environments in the same way we arrive into linguistic environments we might agree there is cause for further investigation.

At the heart of Heidegger's (2013 [1941–2]; 1993 [1954]) critique of technology and technological rationality are the ways in which technical rationality equates being with presence. We touched upon this at the end of Chapter 4 in considering the consequences of *radical transparency* in Benthamite and Posnerite ideas about the value of privacy. The emphasis on transparency is another way of referring to the endeavor to make present and the insistence towards making public. Where traditionally phenomenology tends to be involved with assessment of things that appear for and from human consciousness, Heidegger is interested in that dimension of things that are withdrawn from the world yet deeply influence it. For Heidegger, much of his philosophy is dedicated to arguing that being is *not* presence. To clarify, Heidegger sees the being of things (be this a daffodil, pencil sharpener or suitcase) as never fully present, nor as being in itself (Harman, 2007; Heidegger, 2011 [1962]). This is because for Heidegger a thing is more than appearance, its physicality, usefulness or any means by which we might reduce it down. Rather, as well as the present characteristics (materially I can pick the petals of a daffodil I know to be yellow or metaphorically I know it to be a symbol of Wales), the being of an object is also characterized by a dimension that is beyond what is present (as with the piece of slate). This dimension is the background (personal, social and cultural) context, or what Dreyfus (1991: 10) phrases as 'the intelligibility correlative with our everyday background practices.' Note however that the social and cultural dimension to this is not abstract (coded or reductionist), but refers to the fundamental orienting practices involving the comportment of bodies, meaning and what exists for entities (be these people or organizations). For Heidegger what we take to be present and which we can touch, feel, interpret, utilize, put to work or even sell is only part of the story. For Heidegger, to raise such questions is to inquire upon the most basic question about what a thing is, what it is for an object to 'be' and the more profound question of what being is (a question for Heidegger that has been ignored by Western philosophy). Being for Heidegger is the intelligibility of entities in that being is not a thing or substance, or even process or event, but rather it is the means by which things are comprehended (Dreyfus, 1991).

Being then is not easy to access and for Heidegger this is the point. For Heidegger (2011 [1962]), in *Being and Time*, we have grown accustomed to a public form of scientific representation that is predicted on pure presence, or *enframing*. This is that which has persisted in Western philosophy from both Plato (1993 [360 BC]) and Aristotle's (2008 [350 BC]) *Physics*, the latter's framing of time, and its emphasis on the present, the now and the immediate. This emphasis on presence has given rise to a calculating rationality where all is reduced to Archimedean positions, utility, quantity, efficiency and the logic of amounts, reserves, equipment and stockpiling. These in-order-to discourses emerge in many sites in life (e.g. the need to accumulate, be more productive, produce more cheaply, get faster, go further, be more efficient and so on), and for Heidegger these are underpinned by 'presence-at-hand' or metaphysics of presence. It is to convert not just objects but also subjects into a set of reserves and stockpiles. Such concern about machinism and making technical is not about objects themselves, but rather the unequivocal logic that gives rise to them. For Heidegger technology is not composed of machines, but rather machines serve the process of *enframing* and the tendency towards metaphysics of presence. Indeed, the question concerning the being of technology is an assessment of a modern world that privileges presence and forgets absence. In coping with a sense of time predicated on projection to the future, as demarcated instances or as machinic, we empty experience of being and make ourselves poorer for it. For Heidegger then technical answers to technological problems will not get us anywhere, but rather we need to understand our way out of this milieu along with the consequences of technical rationality.

Decline and disclosure

Heidegger's account is founded on the ways in which the 'being' of technology is revealed and disclosed. Heidegger asserts that throughout Western history and philosophy we have witnessed a decline in our understanding of what it means for something 'to be.' As to what 'to be' means in relation to technology, it is for its essence to be *disclosed*. It is how the phenomenal technicity of an object that is not found in the dimensions or extension of an object (but rather its purpose and character) comes to be. There is within this a pragmatic sensibility because of the emphasis on context, ethnocentricity, the denial of ultimate [Archimedean] perspectives and the ways in which entities become intelligible. This is based on the premise that the technological dimension of our lives both precedes and will supersede us. Like language, we are born into technological environments in which technological items disclose themselves in particular ways. Heidegger's underpinning concern is the way in which technology came to be seen as involving raw materials, reserves and that which is ultimately exploitative. Being, then, is not deontological or eternal but has to do with the forces of history, and that which transcends the immediate and continues to buffer and shape our lives. Heidegger provides a bleak outlook, as for him:

> … the history of metaphysics became the history of the unfolding of *productionist* metaphysics. The technological understanding of being, the view that all things are nothing but raw material for the ceaseless process of production and consumption, is merely the final stage in the history of productionist metaphysics. (Zimmerman, 1990: xv [emphasis in original])

Heidegger's (2013 [1941–2]) concern is less about material events but rather the *modus operandi* that characterizes this driving will and momentum. This is ahistorical, always looking forward and only taking from the past what is expedient to make the present better, faster, progressive, more certain, useful and efficient. This mode is a calculative one and for Heidegger what is lost is human being as it becomes operationalized as the most important raw material for the carrying out of this will. At heart is a question about the cost of making everything present, exchangeable, quantifiable and the organization of life around technical principles and productivity. What is lost in such activity? For Heidegger a great deal, ultimately leading him to National Socialism, because of its claim that it sought to find a new relationship between working and producing that emphasized art, craft and letting-be over alienating working practices. Heidegger asserts:

> The highest danger for the advent of the west is concealed in the fact that the Germans are succumbing to the modern spirit, in they are abetting it along with its unrestricted capacity for organizing and arranging into the most vacuous unconditionality. (2013 [1941–2] §127: 78).

His perverse romanticism and philosophical activism involved a rejection of what he saw as devastation of being, the demise of metaphysics, the establishing of a new order (modernity), distributing and sectioning, making surveyable, turning the world into a causal orderable picture, making mankind machinic, turning humanity into potential and reserve, complete calculation and objectification of the globe, and its conversion into goods and values. The demise and abandonment of being, and philosophies interested in how things come to be for us, represents the loss of metaphysics so to exploit what is present-at-hand. This impetus is never sated and this machinic determinism is not at heart technological, but involves the will to order. Thus where many critical neo-Marxist accounts will point to the liquidation of all to the level of capital exchange, even this isn't Heidegger's critique. Rather, his is that productionist metaphysical impulse beginning in ancient Greece. This metaphysics or technological being is that which miraculates and gives rise to technological arranging of materials, processes and life. The development of devices and technical implements (be these managerial or material) are tools of this will and pressured impulse to be modern.

Heidegger's account of excavational technology helps us situate privacy discussion within a much broader history and generalized will to make selves more present, public and that which may act as a standing-reserve. This productionist tendency of technology is highly one-dimensional in orientation. The twist in Heidegger is that in contrast to hackneyed accounts of modernity, humanity (nevermind a God) is no longer at the wheel. Instead humanity becomes the subject of self-directing technological productionist tendencies. Rather than attributing creation of 'being-technical' to those instrumental in the making-technological of society (be this Sergey Brin, Larry Page, Mark Zuckerberg or those less well-known responsible for data mining innovations), for Heidegger their actions are part of a much more profound disclosure of metaphysical logic going back to Plato and Aristotle. Heidegger's despising of liberalism comes to the fore again because its obsession with the autonomy of the self-positing subject misses a far larger picture. For Heidegger the will to power and its manifestation in techniques, structures, networks and arrangements transcend human will. While heads of digital businesses and those charged with managing state security have argued for the inevitability of changing social mores because technology is changing, such inevitability is part of the metaphysical impulse to make everything unconcealed, available, present and exchangeable. The consequence is that the gestalt of humanity (physical and immaterial) comes to be seen as raw material within an all-embracing totality of technological being. *Dasein* (or that sense of being unique to humans) is made to submit to the techniques and character of instrumental productionist tendencies. Unlike Marxist-inspired critique, in this account no one person or class is in control and although we might decry the behavior of given influential actors or agents, unfolding technological events are part of a much wider and longer narrative that transcends the present. For Marxists, Heidegger was a reactionary thinker who by arguing for a deterministic character of technology is both charged with mystification and guilty of disavowing the possibility of liberation. Both Marx and Heidegger however remain under the influence of Hegel who also saw people as actors in a much larger drama that they themselves do not compose, nor are fully aware of, despite actually producing the apparatus and institutions of the modern era (Zimmerman, 1990). Both too believed that they had uncovered the essence of technology and, certainly in the case of Marx, its relation to capital. They differ however in that Marx's concern is materialism (having upended the Hegelian dialectic) but Heidegger is interested in the longer game and the unfolding of productionist logic that began millennia ago.

A-historical history

Beyond ontologies of presence involving a philosophical and technical shaping of the world to conform to a metaphysical enterprise beginning millennia ago (involving order, causality and calculative discovery), Heidegger's concern is that we have progressed to an age of unhistoricality. For those of us interested in online media and information this has a number of facets to it, not least that the web can be thought of as a flat collective memory. While people have to work hard to recollect the deep and distant past, the most banal of online behavior is readily and equally accessible (and thus quantifiable, cross-referenceable, exchangeable for capital and usable for surveillance and productionist purposes). While this may not be of great interest considered in terms of 5, 10 or possibly even 15 years, when we start to consider that entire mediated lifespans are to be readily accessible by companies and those political organizations with access to their data vaults, the premise of a-historical data mining takes on significance and cause for concern. In Heidegger's account the technological age is 'especially an age of unconditional technology [that] is under way towards unhistoricality' (2013 [1941–2] §295: 231). It is not so much that we are history-less, but that the history is paradoxically deeply shallow and rendered as that which is convenient and distorted because only those dimensions that are capable of being made present-at-hand are recorded. We can identify three parts to *a-historical data mining*: 1) a transparency of history in terms of chronology of what happened when; 2) the development of a flat history where both recent and distant past are equally readily recallable (an important tool for data miners); and 3) where history (in terms of memory), along with human beings, is industrialized by means of transparency and conversion into standing-reserve.

Heidegger's persisting influence

Technology is philosophical, or at least had its path prepared for it by a philosophical tradition dominated by rationalist metaphysics. Where contemporary technology critics such as Lanier (2010) ask whether we are building the future for people or machines, the answer is neither: we are both building and fulfilling a philosophical enterprise that began in Greece. Technology thus involves the objectification of Western rationalism and metaphysical thought, and it is this impetus that connects to contemporary premises of control, connection and prediction. Rationality is that which seeks to break down obfuscation and clarify the nature of relationships by means of empirical observation. While in many settings this is a positive virtue (for example a belief in democratic transparency or understanding the composition of diseases), motives characterized by rationality and making transparent are not always for the benefit of all, as concluded in Chapter 4. Vattimo summarizes Heidegger's position well, generalizing and remarking that 'technology, with its global project aimed at linking all entities on the planet into predictable and controllable causal relationships, represents the most advanced development of metaphysics' (1988: 40). This is a fundamental point as the links to contemporary data mining techniques, the charting of the social graph and intensification of visibility are not just technological and industrial, but finds their roots in philosophy. As mentioned, however, these are not liberal criticisms as Heidegger eschews the subject and its concern with autonomy and self-determination. Vattimo posits:

> The reason for Heidegger's (and Nietzsche's) anti-humanism become ever clearer: the subject, conceived of by humanism as self-consciousness, is simply the correlative of metaphysical Being which is defined in terms of objectivity, that is, in terms of clarity, stability, and unshakeable certainty. (1988: 42)

The point here is that the employment of framing in terms of subject and thereby object is to structure a program where subjectivity is always under erasure and attack. Vattimo continues:

> [...] the subject that supposedly has to be defended from technological dehumanization is itself the very root of this dehumanization, since the kind of subjectivity which is defined strictly as the subject of the object is a pure function of the word of objectivity, and inevitably tends to become itself an object of manipulation. (1988: 46)

This is to say that the very setting-up of subject/object distinctions in Western ontology (consideration of what exists for us) and its tendency towards rational understanding (that which attempts to see, connect and know) is problematic for Heidegger. For Heidegger, taking shelter in anti-technological discourse and humanism is nonsense as the human subject exists as a function of subject/object relations that facilitates technical rational endeavor.

Events

A key task for Heidegger is to reintroduce 'being' as such into questions about life and living. This for Heidegger has been forgotten. All lived activity has a context and rather than look at what things can do and to what ends objects might be put, for Heidegger we should also reflect on the beingness of things that are significant for us. One way of accessing being is by means of *events* and paying attention to the context from which an object, situation or phenomenon is disclosed. Heidegger's usage of this word is quite different (yet loosely related) to my own that I defined at the beginning of this book, so a little caution is required. The definition of events we have been working with so far is that an event is an outcome of a situation that transforms that which is around it. In privacy terms this means that if a privacy event occurs we might trust certain people a little less; a technical system may be overhauled or provided with additional layers of security; national or international legislation may be changed, and so on. For Heidegger (2013 [1941–2]) an event is a phenomenological unity of sorts that allows us to distinguish and recognize the uniqueness and 'whatness' of things. In coming into view and being able to see more clearly 'that it is,' we recognize the dimension of being that transcends objects and can begin to assess the context that gives rise to the suchness and specificness of the phenomenon. If we recollect the slate paper weight mentioned in the introduction to this chapter, we see that it operates within a set of office-based interconnections (desk, lamp, computer, external hard drives, Post-it notes) and is disclosed in reference to books that must not only be read, but whose ideas must also be put to work. The being of the slate comes to be within what Heidegger (1988 [1927]) phrases as an equipmental nexus, or that whole from which the piece of slate is disclosed and finds its 'for-whichness.'

This is a very phenomenological view and we should remember that Husserl and Brentano profoundly influenced Heidegger. Heidegger's book, *The Event*, contains little in the way of undemanding passages on what events are for him, but his description that the event is 'the richness of the simplicity in whose guise the turning of beyng [sic] eventuates while disposing and bestows the showing power of signs' is relatively straightforward (2013 [1941–2] §185: 147). An event then is to be aware of that phenomenological mode of identifying objects and the ways in which they disclose themselves so to 'be' what they 'are.' We will progress to explore the connection of this with behavioral technologies and privacy in the next chapter, but the direction I am heading towards is to see if we can (and should) reconcile the contextual nature of how beingness and events come to pass (disclosure) with behavioral media technologies that seek to provide events for us that have a specific character (e.g. nascent modes of advertising). To provide events for us requires knowledge of the contexts in which we move (traversing websites for example), their significance, and short- and long-term understanding of the subject. It is to understand both the environment in which we move, the subject, and to provide events that resonate with the tone or mood we are in at a given particular temporal moment.

In his radical phenomenology Heidegger (2012 [1936–8]; 2013 [1941–2]) deviates from his teachers by seeing things, objects and occurrences as neither mental projections (Kant/idealism) or physical occurrences (materialism). This is a key point of separation from phenomenology, particularly Husserl (his teacher) and Brentano (Husserl's teacher). There is no straightforward definition of events because Heidegger's interest is deeply anti-reductionist and immune to easy encapsulation (and therefore the Marxist accusation of mystification has some basis). However, Heidegger (2012 [1936–8]: 25) posits that events are an opening to *Dasein*, and what 'opens up in the grounding of Da-sein is the event.' Dasein involves what it is to be in the world, to be immersed it in it, and the means by which all parts of it refer to everything else. Dasein then is to refer to the entire system of human meaning, the morass of connections and our own relationship to this as the being of an entity gains its significance not because of us, but in relation to us. Heidegger phrases the event as a *nexus* and the disclosure of interrelationships of meaning that comprise both an entity under consideration, but more radically the questioner too. An event then is an occurrence of being and for Heidegger we are in danger of losing this sense and level of knowing, perceiving and relating to the world because of the untrammeled quest for objectivity, and making present and immediate. The event is both the recognition of the being of things (that transcends corporeal qualities) and also that mode of beingness, which is being itself. The event, then, is the accessing of the mode of beingness. Put otherwise it is the revealing of how things come to be for us, how they take on meaning and significance, and to latterly dwell on the being of meaning. Heidegger also asks that we reflect on the meaning of meaning itself, or the being of being itself. Heidegger (2012 [1936–8]: 16 [emphasis in original]) posits that the 'goal is *seeking* itself, the seeking after being. Such seeking occurs, and is itself the deepest discovery, when humans decisively become preservers of the truth of being, stewards of that stillness.' For Heidegger such conceptions are difficult for us because of our deep schooling in philosophies and worldviews predicated in presence, immediacy and preference for that which can be readily objectified before us.

A key point here is that this involves an anti-Husserlian argument because the being of entities and situations do not exist because of consciousness, but rather being is disclosed to conscious beings. This involves a degree of separateness and a background by which being discloses or reveals itself. Heidegger's oeuvre clearly involves a rejection of scientific naturalism, but also theory and more usual forms of phenomenology that reduces what exists to appearances in consciousness. An event then, in Heideggian (2013 [1941–2]) discourse, is that being in which unfolding of history takes place. To recognize the being of any event, or the thisness that is the outcome of the unfolding, is to recognize the circumstantial nature of events. These disclosure events are affective and provoke care, in that Heideggerian sense of being concerned, involved or engaged with, rather than meeting things in the world neutrally. Again, bearing in mind Heidegger's phenomenological roots, we do not approach situations and things with bare sensory perception and meet objects as an assemblage of quantifiable parts and attributes, but we *recognize* fences, telephones, lecterns, trees, candles, dumbbells, rugs, books, bathrooms and so on not just by attributes, but by another dimension also. The extent to which machines might engage with such qualitative knowing and perceiving is addressed in the next chapter.

Conclusion

While Heidegger is not the clearest of writers and accusations of obscurantism are not entirely unwarranted, he is of consequence for privacy theorists. His account of enframing, technology and the ways in which the metaphysics of technical being transcend the recent period (and political and economic actors therein) is a valuable observation. Indeed, any critical study that assesses how contemporary communications systems make people more surveyable, transparent, predictable, pliable, understandable, into a reserve, or ready-at-hand, is at least indirectly Heideggerian in orientation. Less well-known is his account of events that this chapter also made use of. While I am interested in Heidegger's understanding of being and that dimension that discloses the character and thisness of objects, my intention in this chapter has been to premeditate and set the context for crimes to be committed against Heidegger in the next chapter. This offers a fuller discussion of the extent to which data mining technologies are able to mingle with being, events and the unique co-created situations that we find ourselves in.

CHAPTER SEVEN

Heidegger (Part 2): On moods and empathic media

This chapter resumes the assessment of Heidegger's unique approach to technology and phenomenology, but in a somewhat delinquent fashion merges these seemingly incommensurable strands. My basic observation is that technology increasingly gives the appearance of understanding the uniqueness of situations, sentiments and the 'thisness' of events as we encounter them in a mediated setting. This leads me to put forward a number of concepts, including: *machinic verisimilitude* that involves semblance of intimate knowledge of people; *dispositional competence* and the passing-off of human understanding benchmarked by predictive capacities; and *empathic media* which is the capacity for machines in an ongoing process to represent and put to use the publically mediated emotional states of people, their intentions, communications and behavioral cues, and to act on them.

Moods

While we might operate under the impression that our sense of consciousness is all-encompassing and alert to what has gone before, what is occurring now and what might lie immediately ahead, this is not the case. Consciousness is a window of, at best, 15 seconds (Donald, 2001; Thrift, 2008). Moods are different and while a mood may feel like something very much present, and it is, it also transcends our immediate 15 seconds' worth of temporal awareness and will probably continue after it. Read in terms of the *mood of information* or that co-created tone of interaction between people and networked technologies (McStay, 2011), this means we may be understood more thoroughly and properly by systems that possess flat and a-historical memories that are attentive to the publicness of ongoing behavior. If we follow Wittgenstein, Ryle and their perspective on behaviorism and what is public (addressed properly in Chapter 10), what we hold as interior, private and dear is very fleeting. Simply put, machines that monitor us and understand our distant past as well as our immediate past behavior have a very good chance of knowing us better than we know ourselves, particularly if we use verisimilitude and predictive capacity as a benchmark of understanding quality. While the question of whether machines undergo Heideggerian events (or experiential specificness) is beyond our ken (although we should not rule it out entirely until we have considered intentionality for machines [discussed in Chapter 9]), for now we can say that machines have the potential to engage and interact with our own phenomenal being, and deal with this in functional terms so to be involved in providing us with experiences that are right for us. Phrased in other Heideggerian language, while machines may never know worldliness (being-in-the-world), they are able to functionally engage with it, represent it, engender experiences for it and also equipmentalize it. On the latter, this is for mediated behavior and communication (along with the myriad of topics and purposes for which this occurs) to be rendered into for-whichness, i.e. productionist logic that might take the form of improving targeted advertising or state security.

This accords with Heidegger's insistence that machines deal in properties that are stripped of significance (that equates to the ways various 'nesses' come to be), yet are able to deftly represent in such a way so to generate experiences that are significant for us. Indeed Heidegger's theory of mind is founded on the fact that cognition is *not* based on the manipulations of representations. His unconcealing of the rationalist tradition means that any technical system said to involve intelligence founded on the symbolization of either external or internal situations, or the manipulation of given representational variables, does not qualify as knowing worldliness. This is because people for Heidegger are not solely rational, or more accurately we do not deal with the world in only a cool detached way, but are involved in a pre-theoretical manner (Winograd and Flores, 1987). Heidegger is right to deny cognitivism and any veridical proposal where algorithms might know being-for-people, although they can certainly engage with it. While one might unequivocally reject the capacity for digital behavioral systems to know ourselves, we must face the fact that behavioral systems increasingly possess capacity for verisimilitude and prediction. (Might Amazon choose better Christmas gifts for ourselves than those people closest to us?) We can dub this *machinic verisimilitude* that can be defined as the competences and dispositions that machines possess that facilitate heterogeneous media experiences that accord with phenomenological flows of everyday life. Clearly theory and practice are not always commensurate (for example behavioral systems may be confused by multiple users of the same device/IP address), but the principle of what I term *dispositional competence* that is tested and benchmarked by the capacity for verisimilitude is correct. While it is unclear if machines will ever undergo Heideggerian phenomenal events and modes of recognizing in the same way that people do, what is clear is that increasingly they are able to interact with human phenomenal being in ways that increasingly feel correct, natural and appropriate.

The battle between John Searle and Daniel Dennett on the possibility of machinic consciousness, artificial intelligence and their dispute about the Chinese Room argument[1] (encapsulated in *An Exchange With Daniel Dennett* in Searle, 1998) simply does not matter for us. This is an argument that has gone on since 1980, when Searle argued that the appearance of consciousness can never be consciousness itself—even if it passed the Turing test, and one were able to carry out a lucid and intelligent conversation with a machine. Once we shift emphasis from intelligence to empathy, the question of whether machines *really* understand us and whether humans have an extra dimension of understanding unreachable by any machinic evolution becomes the wrong question to ask. This is because verisimilitude is able to bridge phenomenological being (the terrain of being, thisness and that character which is aside from material qualities that make up a thing) without making any claims to understanding such metaphysical terrain—whatever the intrinsic nature of human/machinic difference might be. Preempting material to come later in Chapter 10, we might highlight the relationship between Wittgensteinian and Rylean ideas about language and being, moods of information and predictive capacity. In regard to behavioral technology and behaviorism itself, our online behavior acts as a public language that for many stakeholders' intents and purposes, is what we are. This means that selfhood is actually more public than private, which as a proposition begins to shake the impossibility of machinic others knowing the real us. While what must be passed over in silence may exist (to appropriate Wittgenstein, 2007 [1921]), the inferences made about ourselves and our condition by behavioral technologies are significant and not easily cast aside, not least because of their capacities to forecast our interests. Employed on the basis that they are effective, verifiable and have predictive capabilities, they might at least be able to offer a passing-off of what it is to 'know' us (and our 'me-nesses'). This is to simulate connection with heterogeneous interior worlds (that are an illusion, as it turns out) by means of the capacity to predict what we will do. To know another then is to know what they do, have done, who with, will do, where and when. Verisimilitude takes care of the rest.

A question about contemporary technologies we might ask is this: can quantity know quality? Put otherwise, can machines that are ontologically oriented to categories of state know what people discuss in terms of quality? Elsewhere I suggested that processes of behavioral data mining are, in effect, tantamount to mining temporal being (McStay, 2011). This required some caveats to make the point, particularly in terms of how hardware and software processes might approach the distinct sense of what it is 'to be,' found in most accounts of phenomenology. The approach I took was to argue for verisimilitude of understanding, particularly in regard to the capacity to simulate insight and intimacy although the point was not made as forcefully as here. Veracity and capacity can be judged by the extent to which these technologies can pass-off understanding and predict our forthcoming behavior. Using the example of behavioral advertising, I argued that machines are increasingly able to engage with being conceived in the richest of phenomenological terms. This involves the insertion of advertisements into our stream of experience and to be able to balance the need for attention (stand-out from wealth of other incoming data), yet provide a sense of naturalness so to be able to channel and divert experiential streams. I argued that while we might not be able to reach semantic levels of understanding by means of machinic filtering, categorization and the sorting of mediated human states, we can approach this the other way via Bateson (2000 [1972]) who points out that the difference between 'being right' and 'not being wrong' may not be entirely distinct. Indeed, a strict behaviorist (I am not one of these) would argue that there is no machinic verisimilitude—but there is simply understanding demonstrated by predictive capacity.

My approach to moods and information draws on Heidegger's (2011 [1962]) discussion of moods that in turn appropriated Husserl's interest in intentionality (again discussed in depth in Chapter 10). For Heidegger a mood is a state of mind with which we engage the world. Employing common sense we would say that moods are something we impose on the world (smiles are infectious), but Heidegger's point is different in that he suggests that moods are a way of being-in-the-world. They represent an attunement that characterizes being-there and the disclosure of how and what things are. The *mood of information* contributes to the disclosure of what it is to be 'there' (McStay, 2011), and involves the study of why and the specifics of how things are disclosed to us the way they are, and how the beingness of our environment comes to be the way it is. As mentioned, this involves an attunement and the ways in which a 'mood makes manifest' (Heidegger, 2011 [1962]: §134, 173) and how 'the Being of the "there" is disclosed moodwise in its "that it is"' (ibid: §135). There is a public shared dimension to moods that Heidegger would have agreed with (i.e. moods are not private sensations), but he would have found their co-creation with machines more problematic given their incapacity to know worldliness.

Co-evolving authorship

My argument in *The Mood of Information* involves behavioral advertising systems and the ways in which they co-comprise our networked lifeworlds, or those conditions by which the world is experienced (Husserl, 1970 [1936]). More accurately, we are actually part of these co-productive systems. In explaining how we are increasingly co-authors of our own advertising experiences and online media experience, my point here was that behavioral systems might be seen in an autopoietic, ecological and co-productive manner that in part facilitates our lifeworlds so to aid in the production of heterogeneous advertising experiences. The reason for invoking autopoiesis is that it best explains coupling and co-productive processes. As we couple and engage in feedback relationships with technical systems moods of information are co-produced, or the character of an environment produced by both the technical aspects and people-based parts of a system or arrangement. Co-productive relations with information systems engender relationships and modes of production where we may 'act as our own metaphor' (Bateson, 1991 [1972]). Such personalized, self-referential and diverse modes of representation increasingly characterize the delivery of entertainment, media and communicational content. These contemporary technological developments are also expressed well in Kantian (1952 [1790]) inspired critique about natural purpose and self-producing systems that have influenced later autopoietic accounts of digital media (von Neumann, 1963; Hayles, 1999; McStay, 2011). The key point that I wanted to convey was the need for a shift from yet more assessment of advertisements' representational qualities using well-recognized tools of Media and Cultural Studies to make very familiar arguments concerning consumer society, identity and the like, to the assessment of *how* contemporary advertising systems work and to explore their implications. The purpose of this was to shift critical emphasis from the terrain of representation to the domain of nascent feedback arrangements and co-authored experience.

Advertising historically can be seen as authorship by creative teams; the co-creative endeavors of agencies; and going even further we might include market research where a smidgen of our attention time may have been aggregated into the mix that dictates which press, poster, television and other above-the-line advertising we receive (Smythe, 1977; Bermejo, 2009; McStay, 2011a; Fuchs, 2012). However, there was always distance involved and indeed, the postmodern spectacular involved exactly that—spectatorship, and thereby a degree of separateness. Our current situation involves intensification and speeding-up of feedback so to render critical theories of spectatorship less important than assessment of how we contribute to the fashioning of our media environments. As co-creative authors of our own mediated experience it is important we shift critical emphasis to the assessment of *collaborative and co-evolving authorship* between subjects and objects, and to more thoroughly understand the nature of the feedback relations therein. Again, these have autopoietic qualities and that proper sense of cybernetic relations where subject/object or human/machine operate as functional wholes reproducing their arrangements to be able to continue. The invocation of authorship is a useful one, although clearly authorship in this context differs from that liberal and romantic self-willing sense of the term.

While my argument in *The Mood of Information* (2011) was to some extent speculative and many caveats were offered about being in a germinate period of behavioral practices, only a few years later significant developments are being made in technical endeavors seeking to commodify sentiment, mine, fold in and more significantly disrupt the all-too-tidy distinction between phenomenology and experience on the one hand, and the materiality of technical objects and processes on the other. While Heidegger argued there is no 'is-meter' with which to assess 'being' because the existence of being in scientific terms is meaningless (Heidegger cited in Zimmerman, 1990: 224), this relationship between qualitative experience and the technical ways that heterogeneous experience can be facilitated invites renewed interrogation of this claim.

April Fool? The case of Verizon

In May 2011 the US telecommunications company Verizon submitted a very revealing patent application. This provides an extreme but real example of enframing, making present and transparent, productionist metaphysics, reading into how things might 'be' for us, machinic understanding of contexts (or at least verisimilitude) and the nexus of interrelationships that disclose events by means of co-created moods of information. Along with other similar companies, Verizon seek to serve targeted advertising based on what television viewers are doing or saying in front of their sets. While this may appear fanciful and possibly reminiscent of a Philip K. Dick novel, their patent application was for a media system that personalizes content on the basis of who and what is in the room, and what is taking place. Titled 'Methods and Systems for Presenting an Advertisement Associated with an Ambient Action of a Use,' the technology would be capable of triggering tailored advertisements based on whether viewers are eating, playing, cuddling, laughing, singing, fighting, talking or gesturing in front of their sets (US Patent & Trademark Office, 2012). Cuddling then for example may involve advertisements for a trip to Paris. Behavioral indicators included within the application for patent involve a person's size, build, skin color, hair length, facial features, voice tone and accent. It also accounts for pets as well as humans, and objects such as cans of beer, crisps and other objects with which advertising experiences might be tailored. Application 2012/0304206 thus seeks to remedy a failure of past advertising. While advertising is typically generated by what a person is looking at or where that person happens to be (as is the case without outdoor advertising), Verizon has taken the idea of behavior quite literally and seeks to advertise on the basis of what the user is doing, who they are doing it with and on the basis of what else is present in the surveiller's range of attention.

The patent application details a flow chart of how this would work beginning with 'start,' and then progressing to: 'Present a media content program comprising an advertisement break; Detect an interaction between a plurality of users during the presentation of the media program; Identify an advertisement associated with the detected interaction; Present the identified advertisement during the advertisement break; End' (US Patent & Trademark Office, 2012). The application remarks upon the failure of traditional advertising to engage with human (and animal) ambient activity taking place in real time. There are two dimensions to this. The first is that historically, advertising has been quite reliant on human observation and use of focus groups (non real-time human observation). The second reflects a net increase and interest in real-time behavioral monitoring for commercial purposes (via online cookies and possibly deep-packet inspection). The first involves overt visibility while the second is low on the visibility and image-making scale (including other senses). Verizon's proposal combines the two to comprise a behavioral advertising system that does not infer preferences, but rather is able to witness and make true images of what we are doing.

This works by means of a set-top box that has a wide conical range of image detection. Images are not necessarily visual, but include aural and thermal records of 'ambient' behavior going on in our living rooms. Clearly this is a wish list of behavioral data that Verizon would like to gather, with analytical tools including motion capture and analysis technologies, gesture recognition technologies, facial recognition technologies, voice recognition technologies, and acoustic source localization technologies. These are employed to detect actions including movements, motions, gestures, mannerisms of people in a room, location of people, proximity of people to others, people's physical attributes, one or more voice attributes (for example tone, pitch, inflection, language, accent, amplification) associated with persons' voices, physical surroundings (for example one or more physical objects proximate to and/or held by people), and/or any other suitable information associated with the people present. Pets, products, brands, decorative style, objects (for example photographs and pictures) and other animate or non-animate objects are also to be scanned and profiled.

The list does not stop there. The set-top box will determine: which people present are adults or children by means of the physical attributes of the user; the identity of the user (based on the physical and/or voice attributes of profiled persons); a user's mood (based on the user's tone of voice, mannerisms, demeanor), and so on. Should more than one person be in the room, the set-top box will analyze the received data to obtain information associated with each user individually and/or the group of users as a whole. So, for example, if the system detects a user is stressed then it may deliver advertisements for relaxation products. Likewise, if a domestic argument or disagreement is taking place, the system may deliver advertisements for marriage or relationship counseling.

Verizon provides examples of how this would work in practice. The detection/set-top box may sense that a 'user' is singing or humming a song, and by means of a suitable signal processing heuristic identify the name and genre of the song. Also, based on this information, the system may determine that the user is in a particular mood. Accordingly, one or more advertisements may be selected by the system for presentation to the user, possibly configured to target happy people. It will also recognize other ambient actions performed by a user (that may include eating, exercising, laughing, reading, cleaning, playing a musical instrument and so on). For example, if a 42-year-old woman is doing yoga or pilates in the living room with the set-top box on, then the system may deliver advertisements relating to bodily care, activity holidays, spa retreats, products that help relieve muscle pains, healthy foods, and so on. Of course as we watch television today, many of us are engaged with tablet computers and phones, and this is not lost on Verizon's systems. The system may be configured to communicate with the mobile device in order to both serve advertising directly to it and receive data indicating what the user is doing with the mobile device (for example web browsing for techniques to strengthen core muscles, shopping, writing email, reviewing a document or reading an e-book). While delivering advertising on the basis of mood, activity, ambient conversation/group behavior, objects and pets present, the system will also monitor responses to advertisements (for example key words and call to action devices (e.g. 'do this/find us at...').

Confirming the worst suspicions of privacy advocates, the concern is less that this technology will be rolled out in our homes, but rather the meta-tendency and direction of such technical rationality and the will to make present. The example then serves as a repository of Heideggerian concerns about making-present and conversion of lived life into standing-reserve, but also illustrates that being as a totality of context in which a specific person is oriented towards is not entirely foreign to machines. While one may never see Heideggerian 'is-ness' or 'this-ness' meters (although one could connect this with sentiment analysis infographics), machinic verisimilitude provides the next best thing. As such, the question of whether or not technology can *really* know context and selves becomes redundant.

Empathic media

If we are to talk of moods of information, then we are also talking about *empathic machines* and *empathic media*, i.e. those objects and processes able to pick up on the emotional state of others, their intentions, their expressions and actions, along with behavioral cues, and act on them. We might refer to these as *zombie media* as increasingly they are able to answer questions, provide appropriate responses, engage in language games and so on, yet not possess what in common sense terms we refer to as consciousness. The premise of empathic media is to recognize that we are now in a position where we can seriously ask the question about whether machines can appropriately engage with streams of phenomenal experience, and personal and interpersonal contexts, without the caveats that only humans can engage with a variety of 'nesses' (me-ness, thisness or togetherness). Indeed, beyond speculative patents by media companies, machines are already capable of reading facial expressions to track conditions such as confusion, discerning between liking or disliking, and detection of stress or excitement. For example, in research conducted at Massachusetts Institute of Technology (MIT), computers did better detecting a genuine smile than people did (Hoque, 2012).

Note that empathy is quite distinct from sympathy, which is connected to intuition, compassion and 'feeling with.' My intention is not anthropomorphic (or technomorphic) as firstly this will not take us very far and secondly it may well turn out that machines are better at empathy, and comprehending and making use of public signals. Empathy is simply a form of reading social cues and being able to respond appropriately. It is not just a human nicety, but it involves the non-linguistic ways in which we read and understand others. It involves watching, listening, sensing and making inferences about actions or inaction. Memory also plays a role acting as an analytical repository with which to judge (in reference to a set of given values) what is appropriate, and clearly a-historical machinic capacities have advantages over people in this regard. There is much room for disagreement here, not least because empathy exists across a spectrum of academic disciplines (for example psychology, social psychology, media, philosophy, film studies and law). However in all these disciplines, empathy is considered in human and on occasion ethological terms (Coplan and Goldie, 2011), but machinism does not feature. This is not entirely surprising as empathy tends to be framed in terms of morals, ethics, emotions and feelings; and none of these are typically associated with either machines or media platforms.

In philosophy empathy has a number of distinct roots in Hume, Adam Smith (with regard to moral sentiments), Husserl and the hermeneutic tradition (such as Gadamer). For Hume we turn to Book Two of his *Treatise on Human Nature* that discusses the 'propensity we have to sympathize with others, and to receive by communication their inclinations and sentiments, however different from, or even contrary to our own' (1965 [1739]: 316). While we should not confuse sympathy with empathy (the word 'empathy' was not at Hume's disposal at that time) we can overlook this and see that the key is to see empathy as a mode of communication. This for Hume was a means of engendering community, fellow-feeling for others and the contagious nature of moods (be these happy or glum). It was to be judged by effects (or affect), and the ways in which a tone or emotion infused a situation with a degree of 'force and vivacity' (ibid: 317). This does not involve a statement 'I am happy' and therefore believing and responding, but rather it is a set of behavioral engagements predicated on behaviorally expressed public inclinations and sentiment. (At a push one can see it in Heidegger's approach to moods and being-in-the-world). Hume's definition for Coplan and Goldie (2011) involves a form of mirroring, and a sort of low-level of empathy. In our case we shall agree with the public and behavioral dimension to this, but also see it as a form of reading that does not require a response involving imagination, emotion or feeling.

Husserl (2002 [1952]) made use of Herder's notion (2002 [1774]) that we should 'feel into' (*Einfühlung*) an artistic tradition, people or period of history, that itself stems from the Greek word *empatheia*. This is to recognize the uniqueness of any arrangement, to maximize understanding of context and local conditions (e.g. language, geography, history and other social factors), and to imagine conditions of perception and affective sensation. To extrapolate, it is to get as close as possible to the 'nesses,' and events and being, and understand the contexts that gives rise to them. Husserl's approach to empathy is based on interpersonal being, mutual relations, communication and connections between people. It is a bond that involves the co-recognition of ego-subjects, co-presence, the referring back to one another, and an ordered system of indications that comprises continuous experience of/with other people (2002 [1952]: 173).

Empathy involves a co-positing or simultaneity. For Husserl we only reach another person by seeing them as a 'comprehending analogon' (2002 [1952]: 208), but it would be interesting to see what he makes of technology characterized by machinic verisimilitude, and the potential and capacity to engage with human experience. For Husserl, the communal moment has a certain givenness to it characterized by 'a moment of presentification through empathy' (ibid: 209). Empathy for Husserl is inter-subjectivity and while 'feeling-into' is one way of phrasing it, participation works just fine too. As to *what* is being participated in, this for Husserl has an objective quality. A marriage for example is a social experience involving empathy, but this is simply a strong and recognized connection involving (but not restricted to) rich and open flow of information. Again, but in quite a different way, empathy has uniqueness to it as a mode of dissemination, communication and most centrally with which to discern, experience and to an extent inhabit others so to better reach the meaning of actions. My own usage of empathy is very dry by comparison, but still involves affective responses to stimulation, historicity, context, circumstances and examination of reference to others within a social arrangement. My suggestion then is not simply an operational definition where we can say a machine can empathise because it acts like a person, but rather that machines and people empathize in similar ways by means of responses to public behavior. After all, when empathizing, we do not literally feel-into (i.e. enter the skulls of others) but we read and respond to public signals. The merit of the term empathy stems from its predictive power, rather than inference of some quality or essential privatized nature. A weaker version of empathy allows notions of synchrony and co-cognition processes (be this human/human, human/animal or machine/human or machine/animal), but does not assume reciprocity between both parties or some magical capacity to perfectly reconstruct what it is like to be another. Indeed, it is highly doubtful that people actually do this, but instead we represent the supposed experiences of others by means of our own experiences and representational means (just as machines do).

My colder approach to empathy sees it in terms of process, as amoral, as a not necessarily desirable activity (it can be intrusive from people as well as machines) and as a means of reading others. It is predicated on: assessment of states; reading and explaining observable phenomena; hypothesis making (what state are they in?); checking (against known context of individual); testing (reacting to the person being empathized with); feedback (from the person being empathized with); revision (if the first response is not appropriate); and the continuation of the feedback relationship. With as much heart as a dead fish, empathy can formally be summed up as a hypothetico-deductive relationship. Further, empathy is not something that is felt although it will involve the public feelings (behavior) of others. It is not a state like joy or sadness, but a process of being involved with another. This process-based approach sees empathy as distinct from sympathy because where the latter has a benevolent interest in the continued well-being of another (as expressed in Adam Smith's (2011 [1759]) account of moral sentiments), empathic processes do not (despite best efforts of brand consultants for social networks). Sympathy reflects a genuine concern about others and a motivation to improve their condition; empathy is more of an information gathering process. We might add to the differences between sympathy and empathy, in that the latter implies a greater degree of distance and removal from a situation ('I empathize, but...'). Related, torture and cruelty similarly involve the capacity to correctly simulate (if not adopt the perspective of) victims. On this point De Waal (2012) gives the example of a woman being raped in front of her husband. It is both brutal to her yet serves to torment him too by means of exploiting the nature of the connection between them.

In exploring technicity and 'empathy' we should be careful not to get carried away and argue that machines can in some way see or understand our worldview (and the various 'nesses'). While such systems have been constructed to make sense of attitudes, social connections, contexts and public sentiments (behavior), it is clear that machines cannot 'know what it is like to be in our shoes.' The empathic media we are discussing cannot imitate qualia or feelings associated with states and experiences, and do not have analogous experiences with which to mimic what we are perceived to be undergoing. There are then two accounts of empathy to be depicted: the first is the one employed here based on causation and inference; the second is the simulational version, which is disregarded here. This means our experiences cannot be reconstructed and we are alien to such mindreading systems (and increasingly their processes are alien to us). Less formally, such empathic systems do not try to be like us, nor do they attempt to put themselves in our shoes and accredit themselves as having comparable aspirations or wishes. Empathy here refers to something less than simulation, but the net effects may be the same. It refers to inferences made about behavioral dispositions and external stimuli, and the empirical judgments that are made on the basis of these. This is done by means of programmed expertise in symbol manipulation, inputs imputed as representations, manipulation of these according to specified correlational rules, and subsequently the output of further representations.

There are commonalities between people and machines in that empathy in general requires: observation and noticing; inferences; and matching representations of these emotions to internal process of our own. In the case I am putting forward what is critical is the capacity to represent and respond appropriately. While machinic verisimilitude involves alien and functionalized modes of representing and processing of public being, this does not preclude them from being highly empathically effective. This is an account of empathy with emotive and caring undertones removed. The version of empathy presented here connects with Goldman's (2006) account of 'theory-theory' as an approach to empathy, or that expression from developmental psychology that refers to inferences about another person's inner state using theoretical knowledge. It is a theory about people (and in our case machines) that empathize by making rules (or theories) about what they witness so to respond appropriately. This is that process of data accumulation, theorizing, making rules, more data accumulation, revising and theorizing and so on, that applies both to science (theory-theory is also known as the Child Scientist approach) and Bayesian systems where probability estimates for a theory are updated as additional information is acquired (these underpin the logic of data mining). Behavioral enterprises (assemblages of machines and people) are thus those that equate behavior and sets of environmental conditions with mental states and making judgements based on probability as how best to respond.

We should also note the phenomenological dimension to this. When someone (or increasingly something) makes a clear mistake about what we may be undergoing we experience a jarring of sorts, but when empathic processes are functioning correctly there is a dialogic flow (indeed, there may be more than two actors) so for the event to be charged yet pass unreflected upon. In heterogeneous empathic media environments we are able to say that exactly the same conditions should apply if they are to occur optimally and achieve strategic outcomes for web-vendors, advertising networks and so on. Describing machines as empathic does not require we ascribe intelligence, but it does involve a connection, particularly if we are allowed to judge empathy by means of public performance and responding appropriately. A question then emerges on whether we judge empathy by effects or the way in which an operation is carried out. My invocation of empathy is intended to mitigate confusion over intelligence. The latter generally has to do with understanding, reasoning and self-awareness, and while these may be said to be apparent in some of the most advanced artificial intelligence systems, it is something of a stretch to say that processes involved with data mining are intelligent. This is not my interest and I instead argue that instead empathy is a better word to characterize the toing and froing of information, feedback and the loops and relationships within these situations. If we can agree that machines are capable of empathy then we might go one step further and begin to question who is the subject and what is the object in the arrangement. In an empathic situation there will always be an object of empathy and a subject (the empath). In my proposed account of empathic media a person will become an object of the subject's machinic empathic processes and responses.

Conclusion

This second chapter involving Heidegger has paid attention to the nature of processes that allow mediated entities to occur for people. Perhaps bizarrely for some readers, this was driven by an interest in advertising: while media scholars interested in advertising tend to pay attention to the advertising signs themselves, my interest is more infrastructural in how these come to 'be.' This becomes doubly interesting in heterogeneous media environments where our advertising experiences are to a greater or lesser extent unique. The connection with advertising processes comes about because advertising networks have a vested interest in knowing all they possibly can about us. This will to total understanding of context and background begins to appear uncannily as being, or that nexus which may be conceived as the totality of meaning and all possible inter-connections.

In regard to how machines get to know us, this chapter has posited a number of analytical innovations, particularly in regard to empathy. It has not claimed that machines will come to know people as such, but rather machines offer a verisimilitude of knowing. However, I leave open the question of what authentic knowing might be. In making claims about empathic media and verisimilitude, the benchmark for the credibility of this notion is the extent to which they can pass-off understanding, provide content that fits with experiential flows (failure is defined in terms of unwanted jarring), and perhaps most concretely predict our interests and increase purchases and ad-clicks. While the case of Verizon and its patent for total domestic surveillance is unlikely to be implemented in full, the proposal itself is a realization of both threads of this chapter (enframing and machinic engagement with being). It also points to the granularity of engagement that such companies seek to make real and that the age of empathic media has only just begun.

CHAPTER EIGHT

Latour: Raising the profile of immaterial actants

Heidegger and Latour share an interest in technology. Heidegger's approach sees it as part of an unfolding of a much older metaphysics concerned with presence. Latour is also interested in technology's ubiquity, strangeness and even seeming spirituality, and both agree that technology is not just technological. On pervasiveness they also share a common interest, but their means of assessing technology and its function within society are very different. Latour's interest is of a more democratic nature as he seeks to assess technology by means of affect and its capacity to make a difference to social arrangements. The capacity for affect is technology's membership ticket to being a full participant in social life, and for Latour the capacity to affect gives technology a voice of sorts in how society is arranged and governed. This means that in any analysis much greater attention should be paid to the role, affordances and properties of any given technology. In assessing technology and human arrangements Latour sees evidence of one in the other, and two steering concepts in Latour's writings are hybridity and actants. Hybridity refers to an endeavor to find less divisive ways of expressing the relationship between people and the domains in which we live, and the things therein. This also problematizes the idea of pure objects or subjects as both are comprised by significant traces of the other. Latour employs 'actants' as another word for 'actors,' but uses the stranger sounding term to remove traces of anthropomorphism. Moreover, beyond the blurring of objects and subjects, actants might also include sources of action and identifiable processes that make a difference. As will be explored, these ideas about categories have radical consequences, as they involve an assertion about what might be said to truly exist. Actants have efficacy, affordances, produce effects, bring about affects, intervene in arrangements and influence the germination and outcome of events. Critically they are capable of modifying the state of other actors. The broad aim of this chapter is to explore connections between Latour and privacy, and centrally to assess whether it is sensible to consider privacy itself as an actant. This, then, is an ontological and metaphysical account of privacy, by means of the weird and wonderful world of Latour that challenges usual categories of what can be said to properly exist.

Living in the margins

Unlike Heidegger, Latour is neutral on whether technology is good or bad. Given his penchant for science and technology, one can probably assume it interests and excites him. He certainly believes it plays a more critical role in social issues than many of his contemporaries. Technology can be marshaled a number of ways and both aid in regulating privacy as well as exploiting it, as is the case with a Heideggerian reading. Technology may be part of privacy solutions as well as problems (cryptography as a means of factoring anonymity into networked systems is perhaps the most common, as with Tor software). This means there is a strong rationale for understanding more about technology itself as well as the philosophy and miraculating metaphysics that underpin it. While much Heideggerian-inspired discourse centers on the negative aspects of technology (and cybernetics in general) on human life, we might briefly consider the role of privacy-enhancing technologies (PETs) in relation to privacy-invading technologies (PITs), or what Schmidt and Cohen (2013) phrase as technologies that are invasive-by-design (IBD),[1] so to see that technology itself matters.

IBD approaches recognize that the current network infrastructure is predicated on invasion and while some might claim that technology cannot be political, this is to fall prey to a myth of neutrality. This is because all technology possess properties, affordances, tendencies, processes that make possible and deny, and in general make a difference to wider arrangements they are involved in. This is not to say that technology is either good or evil (whatever these may mean) but that because technology is capable of initiating affects within a given arrangement, it takes sides within situations by means of the outcomes it aids in bringing about. Placed then in the context of privacy we might pay much closer attention to the various technologies and processes involved so to give them a fuller role within the assessment of privacy. This does not mean that the technical should dominate, but it does require that objects gain a fuller participatory role in that which is to be assessed. Latour has a pithy expression for this remarking that 'they live on the margins of the social doing most of the work but never allowed to be represented as such' (2005: 73), meaning that in our assessment of human life and societal change, technology tends to be under-represented. In this regard the privacy by design (PbD) literature has much to offer as it argues that privacy can be safeguarded by taking privacy considerations into account from the design inception of systems which process personal data (Lieshout et al., 2011). While potentially over-privileging technology at the expense of other implements in the privacy toolbox (that also contains legislation, regulation, public education programs and financial/reputational incentivization to do more than the bare minimum to comply with privacy law), PbD presents the somewhat un-Heideggerian notion that technology can be used to counter privacy threats. It involves new embedded [privacy] norms at the design stage that recognize the socio-technical nature of systems, the ways subjects and objects recursively shape each other, and critically how systems involving privacy may be imputed with privacy friendly norms and values. Such engineering-based approaches to politics take us some intellectual light years beyond Heidegger as we negotiate with our objects and tools, their affordances, and technical and social histories. This gives rise to what we might designate *technically negotiated privacy*, and the recognition that privacy is not just ethnocentric, but also technocentric (although technocentricism is informed by ethnocentricism and vice versa). On this point about hybridity Callon (2012 [1987]) refers to 'engineer-sociologists,' and the ways in which engineering is intimately connected to social knowledge. Rather than engineering simply being about technical answers to problems—economic, social, political, cultural and social norms (such as privacy) are very much built into the formative stages of technology. Technical knowledge does not stand-alone from social knowledge in the building of technology but (and possibly to the chagrin of some engineers) they are to be seamlessly and heterogeneously considered as a whole.

Porous categories

On considering technology the vast majority of us do not sufficiently explore the technical and material means by which things work. Even fewer of us consider non-material processes that go into shaping technological objects, be these technical expertise, political and legal factors, public influence and market research, economic dimensions, component sourcing, management concerns, and the many meetings and compromises that contribute to hybrid objects (be these bathroom scales or smartphones). The idea of a pure technological 'being' is a difficult one and while we should avoid a reactionary swing to social constructionism, a refresher of its lessons is relevant. While technology might be seen as a means of channeling, and connecting actants and forces of both human and nonhuman sorts, so to create reasonably stable objects and systems (Law, 1992 [1987]), we should be careful of ascribing autonomy. To quote from Latour's *Aramis*: 'For the thing we are looking for is not a human thing, nor is it an inhuman thing. It offers, rather, a continuous passage, a commerce, an *interchange*, between what humans inscribe in it and what it prescribes to humans' (1996: 213 [emphasis in original]).

This is to move into the hybrid domain of Latour's quasi-subjects/objects where the border or barrier between materiality, technology and social processes is much more porous than Heideggerian accounts lead us to believe. This involves understanding not only the affordances of hardware and software, but also other more social elements that contribute to the existence of an actant. As Latour (1996) remarks, to understand a phenomenon we have to understand the multitude of processes that go into its formation, for example non-glamorous but critically important topics such as accounting, management, economics, marketing, engineers, computer programming, scientists, lawyers, regulators, politicians, informational management, usability testers and statistics. While wide-ranging, each influences the outcome of the final assemblage that makes the technological actant neither a pure object nor subject, but a hybrid.

Of subjects and objects, Latour (1993, 2004) asks—why only two domains? In building his account he shifts emphasis away from the subject so to involve the affective reality of objects (defined in terms of their capacity to influence other actors around them). Where social constructionism is interested in human-centric symbolic representations of the world, Latour's approach is an outright repudiation of Kant (1990 [1781]) and his Copernican moment when the accessibility of objects were denied in favor of human projection. Here things as they are in themselves are unknowable, although we are compelled to acknowledge their existence. This is an important point as it underlines the insistence that we should eschew the tendency to textualize all modes of knowledge where we see life as a series of signs, and reality in terms of human projection. While Latour is sometimes lumped with postmodernists, his philosophy is a critique of this excessively introspective and navel-gazing approach that involves a turning inward to our ideas, and the establishing of a dualism absolutely skewed one way that intellectualizes the world, ruptures our connection with it, and loses sight of the most obvious fact that we are of the world (McStay, 2013a). Further, where Kant pushes us towards a conception of freedom utterly contingent on self-reliance and the subject, he neglects to admit the non-human influences that help bring about his proposed freedom.

History and tradition is not just of human-decisions (in themselves conceivably read materially in terms of micro-objects, proteins and amino acids) made in a world of blank objects, but also multiple stories of uncertain entities whose status is not transparent and clear. It is with these we live in association, and it is this flattening or equaling of subjects and objects that is most interesting about Latour.

We can select straightforward examples to make the point. What is a student or academic without books, pens, paper and computers? Our extensions become part of ourselves and if we crash our car into another we admit, 'I crashed into another car.' Similarly, if our computer is assailed by a virus or malware we exclaim, 'I've been hacked' (even if not technically correct). The point of this is that, philosophically, atomism makes no sense as hybrids are everywhere and they have been for a long time. In *Pandora's Hope* Latour (1999) uses the example of guns. While a caricature of a sociological account of gun crime depicts such wrongdoing as 'people killing people' rather than 'guns killing people,' others of a less sociological bent (i.e. materialists) might point out that it is actually guns that kill people because the material base of the event is irreducible to the social qualities of the gunperson. *Because* of the gun, a person is changed from being angry (but possibly normally law-abiding) into a killer. A social approach instead (remember, this is a caricature) sees the gun as a neutral carrier of human will or implement of social structuration. The material approach suggests that our status and nature as a subject is contingent on the tools we hold and utilize. Latour suggests a fusion of subject and object to create a new actant or entity out of the two prior actants, and that this fusion brings about a modification of the original two. For Latour, you are different with a gun in your hand and the gun is different too (in terms of what it is capable of). The outcome of the pairing *transforms* the earlier propositions (set of potentialities) into a new set of potentialities of the hybrid actor. Thus it is the hybrid that kills, not the person or the gun, and such recognition requires for Latour we admit of more actants, actors or agents in society than simply those of a human variety. The wider point is that action is not a property of humans, but actants defined in terms of association. Humans then are in 'commerce' with objects that enable them to act and exist. Everywhere, assemblages and hybrids. We might also detect in Latour a strain of pragmatism, particularly given his tendency to innovate, construct and in general look forward rather than backward. There is also a practical character to Latour by dint of his post-Kantian drive to get things done, and to be able to meet with objects, and assess their contributions by means of affordances and their capacity to make a difference.

Pre-empting complaint that only real actors (people) possess autonomy, deviate, misbehave and are capable of non-cooperation and disagreement, Latour's (2004) rejoinder is that clearly we have not spent enough time in the laboratory where objects do not easily play the roles we attribute to them. The idea that objects are controllable and obedient, and humans are always free and rebellious is a false distinction for Latour. Indeed, for Latour 'what causes beings [of all sorts] to act is still subject to argument' (ibid: 82). All this is to say that we are enmeshed and dependent on things and objects, that objects play a constitutive role in being human, and that they have their own way of acting that impacts on human life. Privacy then is caught up and in part constituted by a field of strategies of people, objects and processes, with namable actants each having their own perspective, agenda, objective and set of constraints (e.g. technical, legal, ethical or financial) about how a process should best happen for it.

Latour's (1996, 2005) account of technology is an approach that sees objects as worthy of respect as well as critical attention. In some contrast to Heidegger he does not see machines and technical objects as expressions of 'being,' manifestations of another realm, or as born from quasi-causation and twilight domains of potential, but rather he sees them as being necessarily immanent and *social*. The word 'social' simply refers to a set of connections encompassing technical items and processes, although Latour points to a more formal sense of representation. His discussion in *Politics of Nature* (2004) involves an extension of democracy or political ecology to nonhumans. The defence of this is somewhat convoluted but the means by how he arrives at this are more important than his proposed republic or parliament of things. This is done by means of disturbance of the proposition that only human actors have will and intention, while objects may only behave and be subject to known causal norms. For Law (2012 [1987]), along with others of an Actor Network Theory (ANT) disposition, this opening-up to the multitude involves the witnessing of heterogeneous elements assimilated into a network. This 'ontology' sees a plurality and coming together of social, economic, political, technical, natural and scientific entities on the same terrain as each other. In *An Inquiry Into Modes of Existence*, Latour (2013) affirms the need to span modes of existence and hybridize symbols/things, culture/nature and the various ways in which these mutually inform and infuse each other so to destabilize any premise of absolute categorical distinctiveness. Where Latour's moderns saw a real natural world with everything else that did not fit jettisoned into the category of 'culture,' Latour (2013) depicts this as a calamitous split and equally, the domain of the symbolic needs to be de-individualized so to encourage the tracing of connections with wider material processes. This is due to the capacity for all sorts of bodies, processes, items of legislation, snippets of computer code, or corporate competition to affectively contribute to structures of how people, objects and actants live.

The usefulness of this for our account of privacy in regard to both propositions made in Chapter 1 on emergent protocol and affect is that this ensures we are examining privacy on flattened and more equal terrain, and more carefully assessing the contribution of a wider range of actants. This means that we can admit a greater set of interactions between different modes of being and the ways in which these mutually affect each other. It is to bridge in any assessment of privacy matters all involved scales (be these security settings, hardware, software, laws, corporate behavior, reactions to surveillance, human or machinic deviations from expected norms, and more) so to ensure we resist privileging any one scale, mode or domain of being. Imbued within this is a deep pragmatism that links well with Rorty's account discussed in Chapter 5. Community and contextual values are not found and imposed, but rather actants as members of a community both have to play by community rules, yet also have a hand in creating these (non-human actants do this by dint of their affordances). The study of privacy in a Latourian fashion is thus the tracing of influence and affect within a specified arrangement, consideration of actants from outside that also influence that arrangement, investigation into the means by which privacy protocol has emerged or been negotiated, and consideration of the consequences for involved actants. This then grants privacy protocol a novel sense of *local realism* as that which is made up of actants and their dispositions, but is also that which transcends them so to form an assemblage.

Meta-stability

Privacy is now within the theoretical terrain of hybrids, transduction and assemblages of people, things and processes. As MacKenzie (2002) highlights in discussion of Gilbert Simondon and transduction, a transductive analysis is to consider processes that generate meta-stability. Put otherwise, this is recognition that stability or equilibrium is temporary because within a system or arrangement there are actants that will eventually disrupt it. This reflects privacy protocol well that may remain stable for some time, but whose norms in the final analysis are finite and subject to change. Underpinning this is Latourian logic where constituents of meta-stable processes may involve physical, biological, social, psychic or technical aspects, whose analysis will always necessitate a preference for ontogenesis (how something becomes a meta-stable event) rather than a focus on what something is. Methodological then, rather than trying to identify privacy in and of itself, a Latourian approach involves an assessment of elements and conditions of influence in the identified assemblage (privacy protocol), with the more rigorous Latourians also tracing the elements of the elements within the assemblage. This is to begin to define privacy itself as an actant. This is an important point for the overall argument being made in this book in regard to a systemic account of privacy because if we are to understand the terms and coming into being of publically recognizable dynamic privacy protocol, this understanding involves comprehension of all actants within the arrangement. Moreover, we should not be blinkered to any particular sort of actants or fall prey to either object/technological determinism or social constructionism, as is often the case. Actants from a range of scales contribute to a *political ecology of privacy*[2] and it is productive to think in terms of a renewed citizenry and expanded state of affect and mutual influence. To point to things, people and ideas working in tandem is to remark upon what is obvious; but to raise the assemblage to actant status with rights and recognized capacity to act in the world is to make a reality claim about what exists. This is a key development. So far we have come to better understand the interdependency of subjects and objects, or what is really a hybrid state of affairs. We also now see more clearly that assessment of dynamic privacy norms requires understanding of all sorts of actants. Our next step is to embrace the existence of immaterial actants.

We can do this by means of Latour who ultimately rejects substance-based views and materialism in favor of the language of assemblages and black-boxes (processes on top of processes). These occur in the here and now, but do not tally with substantialism. If we can agree that process is always primary over substance, this begins to make more sense. Indeed, Latour (1999) goes to some length to criticize the idea of substance insisting we recognize it as that which represents the stability of an assemblage. Substance however is only one form of stability and it is the principle of constructedness that provides *equality* between actants. While difficult to come to grips with, this allows for mutual influence between different sorts of modes (such as obviously constructed things like privacy and Nike trainers, but also an element such as lead). Latour's is a philosophy of process and temporary stability, and the constitution of a given actant may be drawn from all sorts of scales and levels. Further, because of the rejection of substance, a construction may be of any variety yet be quite real. While the point about process over substance is really Whitehead's (explored in Chapter 13), Latour best expresses multi-scalar assemblage-making.

Privacy as a construction and outcome has influence, and affects other processes and associations. Thus for Latour a tree, photograph, swear word, moral, sub-atomic process, religious icon, brand, animal or ecstasy tablet are all actants with equal rights to existence. It is these types of categorical claims that are the most difficult to agree with in Latour, but if we can go along with the premise that an actant is that which is capable of given affects, and influencing and making a difference to actants around it (thus ontologically equating policemen with protons), the same degree of reality can be afforded to all sorts of constructions. This premise is the basis for Latour's (1999, 2013) novel sense of realism that refuses to divide 'artificial' and 'natural.' It allows constructions a greater right to exist by means of recognition of meta-stability across scales. This is a rebuttal of reductionist approaches where things are 'nothing but' [insert chosen scale here: e.g. brain waves, chemical activity, economic practices, etc.] and that search to find substance and some ultimate layer of what things *really are*. Meta-stable assemblages such as privacy, work, marriage, light bulbs and lectures are ontologically real, and defined by their capacity to affect. Life is big enough for Disney and electrons to both really exist. The key point is that just because things are created rather than found, this does not make them any less real, be this privacy, cartoon characters, elements of the periodic table, compact disks, laws, planets or neutrons. The key is in the word 'consequence' and while materialists will rage about the ascribing of reality to many in this list, for Latour realness is delivered because of the capacity for affect and perturbation.

Somewhat reminiscent of the fairy tale Pinocchio, the rationale for this stems from the rejection of substance and that a 'real' actor does not come about through 'nature-sans-artifice,' but through the effects of a given actor on other entities (Harman, 2009: 106). Indeed one only has to look to language for that which is capable of affecting behavior, yet has nothing in the way of spatial, substantial or mechanical extension. Latour's philosophy is a metaphysics of infinite regress and in rejecting ideas about ultimate stuff, Latour brings about a novel non-materialist approach to causation that embraces atomics and popular culture, or 'everything from neutrons to armies of orcs from Mordor' (Harman, 2009: 115). Possibly unfriendly to common sense, what Latour offers is a re-approachment between realism (absolute privileging of objects and objectivity) and idealism (human/Kantian imposed realities) and it is in this zone in-between two imaginary poles that we find Latour amid his quasi-objects, quasi-subjects, hybrids, interchangeable actants or actors, events and black boxes. While this may feel like a deviation from technology and privacy discussion it is not. This is because the bridging of realism and idealism gives credit and power to the premise of rights and thereby privacy. It provides privacy greater claims to being real without having to take recourse to substance, essentialism or foundationalism. Again, this is verified by the capacity for the meta-stable construction to affect and alter that which is inside and around it, which may involve laws, regulation, technology, national security, brand values, or more simply people's behavior towards each other.

Privacy as a black box

As discussed, a key part of Latourian (1987) metaphysics is the rejection of the traditional concept of substance. In *Science in Action* he develops the 'black box' metaphor as a means of critiquing the premise that reality is made of bits, integers or units. In this approach, what passes for stable building blocks are black boxes, or assemblages. It follows too that inside one black box are more black boxes. There is discussion to be had about whether a black box has an identity beyond the elements of which it is constructed, but this would take us off track in this chapter (it is addressed in Chapter 9). Latour's black box is that 'thing' which goes without question. Be this an equation, a social network, a door lock, USB stick, liberty, Marxism or God, it is that which we believe in without question. They are characterized by being a 'durable whole' and the more people buy into black boxes without question, the stronger the box becomes. If people question and disbelieve, the box becomes weaker.

Black boxes are 'machinations that turn a gathering of forces into a whole that may be used to control the behavior of the enrolled groups' (Latour, 1987: 131). They are an assemblage that act as one—or provide a functional and affective appearance of doing so. Reminiscent of Whitehead and his rich account of the nature of events (discussed in Chapter 13) this sees the wholeness of the object dictate its constituent parts. The black box for Latour is not illusory as it exists as a proper construction and actant able to participate in other events, but once opened we quickly discern there is no essence. This might be a photograph, chemical, blanket, doorknob, parsnip or the Apple iMac that sits before me as I write this passage. Critically, too, any black box can be opened up and challenged. Even a parsnip is not resistant to cross-breeding, genetic manipulation and so on. Similarly, chemicals are composed of further black boxes also capable of being unpacked. That which coheres into meta-stable black boxes proves resistant to change, as its associations are strong, although never safe. Whether these are cultural, scientific, political, financial, technological edifices, or most likely a combination of these—they are all prone to ongoing transformations and nascent processes. Black boxes, whether a parsnip or an iMac, are machinic and strategic in that their internal balance of affects keep each other in check so to ensure that it maintains cohesion (Latour, 1987). As a cybernetic metaphor, the black box is that which appears to run itself. We do not question its complexity (a further series of black boxes) but we use it in an unquestioning manner.

Black boxes tend to a form of monopoly, of conquest, a sense of the natural and just the way things are. Their reign however is neither natural nor permanent and whether this is a qualified right, a technology, a nation, Newtonian laws or Manchester United, the veneer of the black box is always open to contestation. Privacy too is an assemblage and black box made up of constituent premises (control, barriers, borders, access, information, body, affect, property and territory). The answer then to 'What *is* privacy?' is the reply 'Your question involves nonsense, but let us explore situations and arrangements where that which we designate privacy comes to be.' The point then is that privacy as a black box and actant is created from all sorts of scales, hence why I have invoked Latour. It is also to reject the idea that it is a thing, substance or that which might occupy a location. It is none of these and instead it is an outcome stable enough to have influence on other actants and thereby be capable of affect.

Privacy as an actant

It is Latour's double-insistence on immanence and metaphysical equality that allows for such a proposition. While difficult to mentally flatten these categories, it does make experiential sense. If we follow Latour, his ontology allows us to speak more forcefully of privacy as an actant because as a co-created outcome it is capable of affect, but because of Latour we do not need to ascribe to the outcome a category of substance, thing or material referent. Privacy as an actant is present when a student knocks on my office door; it exists when we consider it as an artificial barrier to economic growth; and is there if we find ourselves in the horrible situation of having had our body or home violated. Indeed, undergoing an unpleasant privacy event and having protocol breached, such as having one's trusted secrets revealed by someone we considered as a friendly work colleague, outcomes may cohere into protocol to affect future situations within that given social group. On a larger geo-political scale, we can cite one of the 2013 Snowden revelations that the US National Security Agency monitored the German Chancellor Angela Merkel's mobile phone (along with 35 other world leaders). This also involved a breach of protocol and expectations whose outcomes will undoubtedly affect future protocol and inter-relations at a variety of scales, including technical objects, systems, policy, human relations, and the various ways in which these scales interact. Privacy protocol then *is* very much contextual, local and, as will be explored in Chapter 13 on Whitehead, stems from a process-based ontology. As an outcome of events—possibly involving human–human, human–object or object–object relations—privacy is an affective outcome of interaction that may transform the original members of an event. These changes may take the form of laws, regulations, people's behavior, the design of technology, the scripts that inform and regulate behavior of actants within a system, management structures and processes of how data is collected and used. Privacy as an actant mingles with human and technical systems so to contribute to multi-scalar assemblages, among jostling sites and struggles of association comprised of scripts, affordances, wills and multiple hybrid actors. To invoke a favorite word of Latour's, such playing out gives rise to a sociotechnical 'imbroglio.' Critically, too, there is no inevitability of this imbroglio but only associations and the ongoing refashioning of these by emergent actants—with privacy itself being a key actant because as a black box, assemblage and co-constructed outcome it both regulates and, when breached, may modify actors within and that which is around it.

Politics and process

As we progress towards the end of this chapter, I want to look at how Latour and his philosophy is sometimes accused by critical theorists of being indifferent and apolitical because he has limited involvement with the grand canons of sociology. These outlooks see society shaped by processes to an extent external from those being shaped. This is of bearing to the broader arguments being made in this book on privacy protocol, emergent norms and affect because these are less overtly political than many accounts of privacy and technology. To better understand Latour's stance it is useful to go back to Gabriel Tarde—a sociologist who happens to be in vogue, possibly because of Latour himself. Tarde's is an anti-reductionist and very un-Durkheimian view. Whereas much classical sociology tends to view society as more than the sum of its people because of extra dimensions of society that impel, cause, structure or influence, for Tarde there are no supra-human entities that can be turned to for explanation of behavior. This point is key for Latour and Tarde as they reject classical sociology and Durkheim's social facts, or those structures and dimensions that transcend individuals living in a society and constrain them somehow (this is also strongly the case with Marxism). Trading blows with Durkheim, Tarde remarks that Durkheim's view sees [human] society as having a remainder even when people are subtracted; Durkheim counters that life is in the whole, not the parts; with Tarde counter-questioning how these realities came into being if not through human effort (Tarde and Durkheim, 2010 [1903]). For Tarde and Latour we cannot take recourse to transcendent social facts, but we must look towards the field of action where ideas are passed on by individuals, and not to how social groups might collectively affect individuals.

Another influence on Latour is Tarde's idea of society, which is the aforementioned premise of society as multi-scalar that includes all sorts of non-human entities (be these bacterial, celestial or machinic). For both there is no distinction between the natural and social domains, but rather everything from microbes to markets is a society. This invitation to non-human processes and objects is an important one if we are to properly accept the social value of privacy as an emergent and affective de-essentialized assemblage. This is abstract and 'society' may feel like a strange word to use, but if we exchange this for 'ecology' it makes more sense as ecology involves connections between actants, process and emergent behavior (in our case privacy protocol). If we put the two together and make a temporary portmanteau 'social-ecology,' this possibly becomes even more palatable. This better allows us to understand the mutual influencing and co-emergent behavior of actants of all sorts, across the full variety of scales, that impact on any given situation ripe for analysis. Importantly too, with a little imagination, we better recognize that people do not sit at the center of our expanded society (because there is no center). With an expanded view of social systems, we better recognize that perturbations at distant scales can deeply affect our more immediate worldview.

Latour's (2005) self-appointed task in *Reassembling the Social* is to shift sociology from being the science of the social to being the science of association, and to disallow the social determinist premise of hidden forces ('power,' for example, being a favorite of critical theory). This participates in the aforementioned relationship-from-a-distance that Latour and Rorty enjoy, as Rorty (and pragmatism more broadly) similarly critiques the premise of underlying forces that might determine human history and communities (see Rorty, 1998: 228). Indeed, Rorty asserts that supposed hidden forces are simply knowledge claims that keep intellectuals in business. At the far end of this type of writing, most of us will be familiar with the worst excesses of critical academic discourse that is akin to conspiracy theory (sinister unverifiable actors at work influencing situations). Academically, there is also the very real possibility that unscrutinized application of theory to situations may frame, distort, close down or obscure inquiry. This is not to suggest that past associations and relations cannot influence the present, but rather that the existence of influence has to be traced and demonstrated rather than uncritically be inferred to exist. This means it should emerge from the factors at play within the situation being assessed and not be assumed *a priori* to exist and determine, without being accountable for precisely how a past situation influences a current one. The key means of testing this is to trace association. This requires an altering of perspective as what is social is not a given, but instead emerges out of the tracing of new associations within an assemblage. This then is to reject the premise that social events are manifestations of larger hidden and generally sinister operations that structure and organize life. This leaves Latour (and the systemic account of privacy presented here) open to accusations of being apathetic or uninterested in politics. Latour's answer is to highlight that:

> ... power, like society, is the final result of a process and not a reservoir, a stock, or a capital that will automatically provide an explanation. Power and domination have to be produced, made up, composed. Asymmetries exist, yes, but where do they come from and what are they made out of? (2005: 64)

Latour (1999) sees nature itself as a political process, and as a complex articulation of human and non-human actors. It is not the case that Latour is apolitical, but rather that he will not allow nature (understood in immanent terms) to be dictated by macrosociology, or by explanations from elsewhere. This seems reasonable and another bug-bear for Latour is that we have allowed ourselves to be mastered by ideas that suggest anonymous fields of discourse that make us act. As Bijker and Law, Latour's fellow socio-technology actors, baldly put it: 'There is no grand plan to history—no economic, technical, psychological, or social "last instance" that drives historical change' (1992: 8–9). Along with Tarde, Latour urges we should focus rather than broaden our perspective and rather than take recourse to expansive theory, we should focus and pay renewed attention to the detailed workings of assemblages, their multi-scalar composition, histories, vectors and trajectories, movements, associations and connections, and events that come to constitute actants.

Conclusion

The Latourian reaction to Heidegger presented in this chapter ensures that technologies are de-automated, but also grants them individual rights. It makes technologies fully-fledged actors because of their affordances, capacity to act and ability to influence. This is an important point for any discussion of dynamic or contextual arrangements. Moreover, Latourian argument about a flat ontology in which substance is devalued in favor of process facilitates the existence of a greater range of constructions. By de-substantialization and defining the strength of a construction by its capacity to affect, this provides privacy a mode of existence. What matters is that in understanding privacy as both a black box, outcome and contributor to further processes, we recognize and study how privacy affects that which is around it (for example its affects on laws and policy, flow and modulation of information, use of third-parties, where one keeps personal letters, hardware and software design, interior and exterior architecture, or management board meetings by social media companies on corporate strategy and winning trust). Latour then helps us to establish the ontological rights of privacy as an affective assemblage that comes to be by means of interaction and events. In many ways we have taken a necessarily long way around to simply say that privacy is a social norm. However, in taking this route we have a much better grasp of what the social consists of, play between scales, the range of actors potentially involved, and that such norms are emergent and do not sit behind the scenes determining. Privacy is not a stable unchanging norm and changes to protocols emerge through 'trials of strength,' the making of new associations, solidification of existing ones and the weakening of others.

CHAPTER NINE

Phenomenology: The rise of intentional machines

At heart, phenomenology is about experience, awareness, how things come to be for us, and detailing subjective goings-on in objective terms. It is keenly sensitive to suggestions about the constructedness of the world-of-our-experience, and the ways in which perceptual faculties and modes of understanding play roles in structuring our lifeworlds. The purpose of this short chapter is to inquire upon whether it is sensible to ask these questions about non-humans and media technologies, particularly those of a surveillant variety. I argue this question is valuable for privacy matters because the answer offers insights into the relationship between machines, attentiveness and people. Of course, attentiveness not only involves machine–human relations, but also those of a machine–machine sort. In progressing this, I develop an argument that began in Chapter 7 on empathy, and assess the [philosophically] controversial premise that machines and empathic media may have intentional characteristics. That is to say, they are able to objectify and intend towards things. Thus, while intentionality as developed by Brentano (1995 [1874]) and Husserl (1970 [1900]) is keenly limited to people (animals receive a brief mention, but objects not at all), I argue that machines are increasingly intentional.

Intentional machines

As an approach to generating understanding, phenomenology describes rather than explains or analyzes. This is to open up and bring to the fore that which was always there, but passes unnoticed. It is interested in the lived world and deals in descriptions of the world and experience without taking into account psychological origins or causal explanations, as evinced by sociologists, historians or scientists. It is not that these areas of knowledge are denied, but rather that for Merleau-Ponty (2002 [1945]) they are somewhat dishonest and naïve by refusing to acknowledge consciousness and the lived aspects of experience and life. This is also Husserlian and subsequently Heideggerian where to be conscious of something is to be engaged with a specific reality recognized through definite experience (Harman, 2005). To recognize an object is not to point to an assemblage of properties, but to recognize an experiential form that supersedes but includes these.

The notion of intentionality itself derives from Brentano (1995 [1874]) and his famous passage (pp. 88–91) on what exists in the psychological realm. This is what Brentano refers to as the *inexistent* objectified part of a presentation to which our actions are directed. It is to recognize that each mental phenomena 'includes something as object within itself' (ibid: 88). Note too that inexistence does not equate to non or lack of existence, but rather to the essential part of a presentation (mental phenomena) that involves orientation to some object or unity. These mental phenomena are wide-ranging and, in addition to the world of things, include judgments, recollections, ideas, expectations, inferences, opinions, doubts and feelings. These all refer to an object of mental sorts that exists within the presentation in mind. This object is a unity, despite being comprised of multiplicity. There is then a givenness of the thing within our mental presentation to which we orient our attention. Brentano posits that intentional in-existence is exclusive to mental phenomena and that no physical phenomena exhibit anything like it. I disagree and argue that intentionality is much broader.

Intentionality is the presence of a thing to which our attention is drawn. Brentano tells us that this thing does not have to be an object and this caveat is an important one for my account. Objects themselves are never whole because we only experience them partially (if only by seeing one side at a time). Rather we experience or objectify objects by means of making a *principle* of our experience of them. Clearly the influence of Kantian (1990 [1781]) idealism, representation, human projection and phenomenological creation are at play here. On objects and things, the coffee mug on my table is of given weight and dimensions (probably larger than most!), able to hold a specified volume of liquid and is resistant to high temperatures (relative-to-my-lips), but it is not these properties that are present for me. Rather the mug has a presence (or its mugness) that transcends in my mind these characteristics and it is to this my mind is drawn, or 'intended towards.' However, Brentano's account is broader than this and an object, as mentioned, does not necessarily mean a thing. This is to allow through the backdoor all sorts of other processes and entities without fixed corporeal form. Intention, then, is that which structures experience (Husserl, 1970 [1936]) and how things appear to us so to provide an experiential unity (Merleau-Ponty (2002 [1945]). Harman (2005), clearly influenced by Latour, makes a key move for phenomenology and grants intentionality to objects too so 'the concreteness of intentionality will turn out to belong to *every possible layer of reality*, not only to human awareness' (2005: 23 [emphasis in original]). This is a very Whiteheadian move that shifts us from phenomenology's neo-Kantian insistence on the sovereignty of human-access to a place where, *objects objectify each other*. This is a point to be underlined as we move the process of intentionality from being a solely human activity to being a process that both objects and creatures engage in. Bearing in mind phenomenology's interest in how things come to be, this renewed sense of intentionality becomes cosmological as it applies to everything.

We should add a caveat too that there is nothing innately correct about intentional processes. False positives, incorrect processing of signals, askew perception and environmental conditions may all cause problems for intentionality. Dennett (1998 [1987]) makes a distinction between intrinsic and derived intentionality. Intrinsic intentionality involves intentional processes that are self-generated (as we ascribe to people). Conversely, derived intentionality is found in machines designed to recognize coins or patterns of information, so is provided by a designer. This seems to be a simple split of authentic/inauthentic and human/machine. Dennett however asks us to apply the rules we ascribe to derived intentionality and machines to ourselves. Are there really intentional deeper facts that apply to us but not machines? Do people have access to a more ultimate and objective reality than machines that may be tricked into accepting a variety of things, whether this be the use of Euros in pound-coin slots, or computer-based Trojan horses that access operating systems by appearing to perform desirable functions? The answer is that we make mistakes too, or more accurately each filter stage is not granular enough for the task in hand (and therefore mistakes are made). We can be provoked into believing that we are engaging with one thing when we are not, and our cognitive systems can be confused. The point of this passage is to recognize that if the proposition of machinic intentionality is to be critiqued as a proposition it cannot be on the ground that we have access to some 'other' reality. Instead, although we have fancier intentional processes at our disposal, these are also capable of being duped in a similar way to machines.

Some examples might help here. To return back firmly to privacy, we see this in behavioral advertising systems as *machinic modes of intentionality* structures experience for them. They have a towards-which modus operandi, are attentive *to* something of a particular sort and make principles of their object of attention (e.g. packets of data with a specified header address or payload). One might rightly say, 'well, they've been programmed for that' and one would not be wrong. However, all intentionality takes place against a background of persuasions and capacities that both enable and contribute to directing intentionality (Searle, 1983). Intentionality is far from being the preserve of humans as at all levels *there is that which structures how things appear*, how things are and what is to be for something. As such, people do not have exclusive rights to principle-making, feeling-out and objectification. Note too that awareness and cognition is not a condition of intentionality, but that this structuring principle is better framed in terms of Whiteheadian (1985 [1929]) notions of prehending. Discussed in greater depth in Chapter 13 this involves 'prehensions' and the ways in which elements might be connected with one another so to form an entity or event. Prehensions thus involve ways in which objects experience other objects, perceive them, feel them, interact or take them into account. This is sensual, possibly anthropomorphically so, but Whitehead's reasoning for being able to apply this to non-conscious beings (or 'occasions') is the way that one occasion responds to another. This is object-oriented phenomenology, and Harman drawing on Lingis more poetically remarks that:

> ... [objects] encounter one another not as stupid inanimate bulks working with mechanical torpor, but as topological bulges in the world, as imperative objects never fully manifest to each other but communicating with one another through the levels that bring their qualities into communion. (Harman, 2005: 68)

This is less about perception or cognition as with those accounts inspired by Husserl or Merleau-Ponty, but a theory of 'the translation of forces between objects' (ibid: 77). This is what might otherwise be discussed as an affective account and refers to situations 'where our bodies meet with the voluptuous textures of entities' (ibid: 3). Such an affective or carnal account of phenomenology recognizes that what we meet in perception are neither objects themselves or raw simple qualities (the object in full never manifests but subjective qualities we experience are always connected to an object), so we find ourselves mid-way, or in what Harman articulates as a 'carnal ether' (ibid: 4). This is a means of finding a way between the caricatured poles of science (and its dodgem-like causal approach) and phenomenology (with its post-Kantian interest in how things appear and recoil at physicalist explanations). The appeal to the carnality of objects, processes, systems and non-human phenomena is a rapprochement that must come about as theory and philosophy will not serve much use to anyone locked in a castle of its own making away from anything non-human (as tends to be the case with philosophy established on the phenomenological tradition).

Developing the point on carnality Harman, in discussion of Lingis (1998), shifts the emphasis away from human experience to other levels and sites where carnality takes place. This, for Lingis, takes the form of understanding the imperative of things, a de-centering of the sovereign Kantian subject and of understanding how other phenomena affect us. For those not familiar with Kant, this should appear straightforward; for those familiar with Kantian problems of inaccessibility to things and the world, this will appear controversial. An object-oriented phenomenology has a number of implications for discussion about privacy, particularly in regard to the status and nature of what is involved in the processes of machinic attentiveness, surveillance, privacy and knowledge-based events. By admitting objects into the fold and ascribing them with intentional capabilities, we allow for the fact that objects may be capable of empathy, albeit by very foreign-to-us intentional means. Moreover, by recognizing intentionality and empathy in things other than human, we raise their status and possibly are slightly humbled as we reflect on the human propensity to believe that they have sovereign rights to intentional processes.

If phenomenology is to be at all relevant there is a need to recognize that historically objects are not even granted less-than-stage-prop-status to life and perceptual events, or worse possibly even as not properly existing. Instead by means of ascribing intentionality to objects we move towards the view that agrees it is increasingly important we pay attention to the processes of watching, objectification and the principles of attention by which machinic intentionality operates. By decentering intentionality from people and expressing it in micro and more carnal terms (and thereby disconnecting it from 'intelligence') we renew our understanding of generalized attention and intentionality throughout the cosmos. Further, we better recognize the intensification and complexification of intentional processes taking place by nascent surveillant machines.

Conclusion

This chapter has inquired into the relationship between machines and phenomenology and the extent to which machines can be said to have in-existent interests, or that to which attention is oriented. This was asked so to make the observation that being intentional and having interests are not especially human. By means of a brief foray into object-oriented phenomenology, we might now better appreciate the nature of surveillant machines, watching and machinic intentionality.

CHAPTER TEN

The subject: Caring for what is public

The constitution of subjectivity is a question that has haunted professional philosophers, post-structural social theorists and teenagers engaging for the first time with the mind-bending question: what is a self, and what or who is a subject? Critchley (1999) points out that linguistically it comes from the Latin *subjectum* or 'that which is thrown under.' Seen this way the subject is an underlying support or fundamental stratum for other qualities. More generally, it is that which persists through change. Classically it is conceived in metaphysical terms so for Aristotle's *hupokeimenon* to be a foundation, or a grounding principle by which an entity is discerned and endures. In Aristotle's (1998 [350 BC]) *Metaphysics* this is the positing of an original subject, a substance of sorts, or a primary that is not a predicate of anything else. This to counter Heraclitus and his argument that the world (and the self) is in constant flux and that what seems permanent is also subject to flux. This view of the persisting subject is also a constitutive one that begins in Descartes, his separation of mind from bodily materiality, the distancing from the world that provides the subject objectivity and reason, and which arguably reaches its highpoint in that neo-Kantian transcendental sense of a self that accompanies all our representations. This continues into traditional modes of phenomenology, the 'subjective turn,' the ways in which the subject plays a role in constituting objects for its self, and the emphasis on experience and *res cogitans*, or that mental and thinking substance which constitutes a domain called 'Mind.'

If we are to allow for a self that endures, where is it and in regard to privacy, how does this connect to more mainstream liberal discourse on the autonomous subject as established in Chapter 3? To summarize, the traditional view of the subject conceived within classical liberal discourse involves will, agency, continuity, stability, sovereignty and that private sense of I [think, believe, feel, do, disagree, am, at], although this has come under sustained and correct critique on the extent to which the subject is self-positing. Two lines of interrogation can be identified, firstly from that Nietzsche-Heidegger-Derrida continuum and questions on the extent to which selfhood is contingent upon language, social and political structures. The second is more fleshly and asks about the philosophical construction of selfhood when a human being is read not as a discrete entity, but an assemblage or society of neural, cardial, bacterial, parasitical, sensational and electrical forms and scales (to name only a few). This is Harman's (2010: 41) observation who in developing this points out that identity and selfhood are reliant on all sorts of different types of cells and independent organisms living inside of us. Moreover, the specifics of this physicality exist beyond most people's understanding and control. As Harman (ibid: 86) later comments: 'The empirical fact that I happen to know my own secret thoughts better than the center of an uncut watermelon does not amount to a valid ontological privilege for thinking or language.' To further illustrate this point, one need only consider the effects of brain disorders or even the slightest of chemical imbalances for our decision-making processes. However, despite self-deconstructive practice and much-needed paring back of egoistic pretentions, the passing of the subject has also been overplayed. We know this because of the demands that others make of *me*, and to deny this is to raise the possibility that we are not responsible for our own decisions (also see Frank, 1989; Critchley, 1996). Despite the fact that the immediate timeframe of consciousness in which we can experience that sense of I disclosed in a temporal frame is relatively short, there is a self that transcends these periods of reflection that possesses tendencies, character and definition. This lends stability in that we do not create ourselves anew each moment, but memories, knowledge, habits and traits persist and develop. This chapter is less interested in political questions about post-structural subjectivity (is self-responsibility diminished because of the constructive influence of discourse and language?), but rather *where* the self exists. Addressing subjectivity this way, it attends to premises of interior/exterior selfhood and identity in relation to language, and debates over public/private distinctions. The question of the location of the subject is important because it has bearing on whether there are interior selves to be understood. If we are more public than we first thought, this also has implications for what behavioral systems may come to know about people by means of intentional machines and empathic media. Simply put, if subjectivity is to be found outside of our skulls rather than inside of it, might we care a little more about what is public?

Identity, expectation and moods

Increasingly we find evidence of ourselves in advertisements, marketing offers, news updates and other online situations that involve rich modes of heterogeneous feedback and real-time metaphors of our own interests. Much of what passes for being 'us' is predicated on probability and expectation rather than by means of something more innate. We very rarely know something for sure about another person but we rely on degrees of probability. For example, my line manager assumes I am male. We have never shared a urinal together but other indicators assure her I am what passes for male (I am sometimes unshaven, I have a deeper voice than most females, I do not possess breasts and on applying for my job post I indicated I am male). We live then in a matrix of expectation rather than certainty and these come to be through sets of traces. This is the subject as expectation whose veracity and authenticity is judged by the capacity to deliver on our expectations. The same occurs in an online setting in that the nature of our subjectivity is de-essentialized and inferred by means of behavior, signification, networks of association and algorithms, or those points plotted either on an individual plane or within several planes. The subject seen this way is a representation of given variables and ingredients of particular proportions drawn from similar or multiple dimensions and domains. As Vattimo (1988) points out, being has much less to do with stable properties and structures but rather is a process or event. It occurs when being 'historicizes itself and when we historicize ourselves' (1988: 3). In regard to the subject and the online environment this means we do not have to concern ourselves with the shibboleth of an authentic interior, and if being comes to be because of historical contingency and interacting situations/events, this is something that data mining systems can work with in their hunt for being. This is woolly, but it is telling that Vattimo uses the language of 'at what point' Being itself is' (ibid). Read otherwise, this emphasis on points and intersections is a geometric reading of discourse and being, and as soon as we read events and situations this way we are in the terrain of algorithmic conceptions. Discourse is thus reduced to a continually developing arrangement of coordinates, and a 'point' is simply a representation of an intersection of multiple planes and dimensions (this might involve purchasing patterns, health queries, types of people connected to on social networks and other mediated areas of our lives). The point then is that the subject comes to be by means of expectation and rather than discerning the subject by means of special interior access, it is simply a shorthand convenience to handle and categorize through public and behavioral means.

As Hayles (2005: 62 [emphasis in original]) phrases it, where once we turned our attention to bodies of texts, we now increasingly pay attention to 'bodies *within* texts,' and the means by which we are mediated therein. We do not understand digital selves as static entities but rather as that which comes to functionally be by dint of interaction with sites and cookies as we traverse and move across the networks. What passes for the self is held together as an assemblage of traces of expectation originally located outside of the self. The premise that our public traces are readily converted into useable languages sits very well with discussion of databases and the reconfigurability of the public subject for specified purposes. Poster makes this point best, remarking that such informational arrangements enact:

> ... a radical reconfiguration of language, one which constitutes subjects outside the pattern of the rational, autonomous individual. This familiar modern subject is displaced by the mode of information in favor of one that is multiplied, disseminated and decentered, continuously interpellated as an unstable identity. (1995: 57)

Poster's still radical but accurate post-humanist point is that the subject is linguistically constructed in databases and subject to surveillance and informational commodification. To make the broader point: ultimately the subject is rendered into textual form and decisions are made about data subjects' behavior, whether these be flagging up for closer surveillance and possibly arrest, or the targeting of advertisements for Marks and Spencer's underwear. Indeed, this is a key area of inquiry for surveillance studies that employs the term 'data-double' to refer to our digital proxy selves or data assemblages (Hier, 2003). While a useful notion that has been accepted at policy-making level, we might also pay attention to what is phrased here as *body-doubles* and the very real need to use proxy selves to affect and influence corporeal bodies (be this for marketing or security purposes).

The direction of my argument in this chapter however is that rather than simply noticing that we are more public than first thought, we should also recognize the need for greater care of our de-centered and de-sovereigned selves, and the extent to which affective parts and traces are found within systems external to ourselves. We might find the cybercultural expression 'post-human' disagreeable and difficult because of the 'after' indication. This is particularly the case for those of us interested in privacy that is founded on liberal humanism (agency, will, autonomy, freedom) that post-humanism questions in light of machines and skepticism about the sovereign role of people in the world. Another reason to be uncomfortable with this is because we have always lived more publically than liberal conceptions of the self/human suggest and to ascribe the 'post' prefix is to miss this basic observation. Indeed, to return to Bentham, he tells us that that 'Actions are a tolerably adequate interpretation of sentiments, when observation has furnished us with the key' (1834: 101). Post-humanism is exactly that—that which comes after humanism. It is not a generalized condition but a reaction to a particularly set of humanist propositions and, as Bentham indicates, other philosophical schools have for some time been confortable with the subject living publically.

Access to others

If behavioral systems and those involved with sentiment analysis may be considered privacy-invasive, to what extent can they be said to access our *real* selves? This is to inquire upon some basic but difficult questions: can we really know the private lives of others; and what are the conditions by which this might be possible, if at all? In journals dedicated to philosophy (journals surveyed include *Journal of Philosophy*, *Mind*, *Nous*, *Philosophical Quarterly*, *Philosophical Review*, *Philosophical Studies* and *Philosophical Research*), the dominant approach to privacy concerns questions of the mental existence of others and the possibility of others having sensations, feelings and thoughts akin to or the same as our own. Are experiences of given phenomena common or is all experience private? To a degree the answer to the latter must be yes for the simple reason that *that* person is having *that* experience, rather than another. A question remains though over whether an experience of a given object or situation is unique or common. If the latter it can be said to be potentially public (knowable by others), but if it is unique it remains always private (unknowable by others). We have no absolute guarantees on this matter but in general err to the idea that what we experience is shared and knowable. If there are intrinsic properties knowable only to our selves then this is only available by 'privileged access.'

Dispositions

The notion that we have special access to an interior self is one that is rejected by Ryle (2000 [1949]) who repudiates the common sense view that emotions are 'something' we experience. Ryle instead reduces thoughts and emotions to mere 'dispositions' of a body to behave in certain ways under certain circumstances. As will be developed, the notion of public dispositions (as opposed to private/privileged access) is an important one. Ryle's is a deeply anti-Cartesian position and his argument is a thoroughgoing rejection of the double-essence or popular dualism approach. This is where we have a non-extended thinking thing (mind/nonphysical) inside an extended non-thinking thing (body/brain/matter/material).

For Descartes and other dualists these cannot be identical with each other, and therefore the mind is non-substantial yet occupies spatial properties and is in intimate contact with the brain. For Ryle when we talk of qualities of mind we are not discussing some form of 'occult' episode of which acts and speech are effects, but instead we are referring to those acts and speech utterances themselves. This is because for Ryle there cannot be a 'ghost in the machine,' but only dispositions. His view is a denial of introspection and an affirmation of behaviorism that sees the mind as not something behind the behavior of the body, but rather as part of the behavior of the body. This includes that which is *publically* measurable and recordable, and might involve movements, utterances, noises, temperature, chemicals released, interaction with environment and so on. In a mediated environment this will involve greater emphasis on assessing choices, utterances/shared content and connections with other public actants.

While behaviorism remains deeply unfashionable and is not even tabled for discussion within my own background discipline of communication, media and cultural theory, there are lessons to be obtained. Skinner puts it in the following terms:

> Human thought is human behavior. The history of human thought is what people have said and done. Mathematical symbols are the products of written and spoken verbal behavior, and the concepts and relationships of which they are symbols are in the environment. Thinking has the dimensions of behavior, not of a fancied inner process which finds expression in behavior. (1976: 130)

This is a rejection of the privileged access or privacy argument, and instead we only have access to the evidence provided by our own actions. The key part of this argument is that mental life is an invention that causes problems when we seek a factual diagnosis of how we work. Introspection does not give us a sense of how things really are inside our skulls and while it may provide representations, it does not grant a privileged perspective on what is going on inside. On interiority and mentalism we are also left to question what precisely it is we might know? If we stop, dwell, focus on an object, perhaps even pinch ourselves, possibly even jump up and down, or go for a mountain bike ride, we might try to mark the point an intrinsic occurrence happens. What precisely do we 'know' or experience that warrants the label of private experience? For Ryle, to talk of mind is to talk of a capacity to do certain things, but not to talk of something that strictly speaking exists. There is in this strict view of behaviorism no relationship between mind and body, because there is nothing more than the brain, capacities and dispositions. This sees a rejection of an interior theater of emotions, of inner experience, and only the granting of public behavior. Indeed, the idea of 'inner' for Ryle is tantamount to mysticism, which in itself has consequences for border-based conceptions of privacy. This is not a denial of feeling or emotion but rather Ryle's endeavor is to highlight inner states as fictitious propositions and entities. We do not have a feeling and then behave, but this all bonds together (one could say as an assemblage) in terms of being public and brain behavior. A feeling, then, is a propensity to behave a certain way. This begins to blur the idea of private inner life and public outward life. Hartnack provides the example of vanity, remarking:

> To be vain, for instance, is not to have special experiences or feelings called 'vanity feelings' but is to have a propensity to behave in a special way, namely to be apt to speak about one's own merits and to be silent about other people's merits and to do many other things, in short all the things we expect a vain man to do on the basis of which we judge him to be vain. (1952: 411)

We might already see a hint of the argument being developed: intentional and empathic media do not need to read interior minds but instead sentiment and character are very much public dispositions—they come to be through repertoires of behavior. This begins to up open up some interesting ethical and legal terrains. If personality is not private property but is by definition that which is public (because behavior is public), where does this leave us? How can we account for protecting privacy when what was thought to be private is actually public? The answer to this is again that we should not think of privacy as seclusion, but rather as the modulation and management of contexts, protocols and connections. The premise of an isolated private individual has been over-played.

Being public

While the liberal and sovereign self protects borders with access-control faculties, what is actually guarded is already far more public than first thought. For example an article by Kosinki et al. (2012) on predictability, traits and attributes shows that Facebook Likes can be used to accurately forecast sexual orientation, ethnicity, religious and political views, personality traits, intelligence, happiness, use of addictive substances, parental separation, age and gender. After all, borders are difficult to manage if the interior is really outside. Not a perfect metaphor, but it is akin to emptying one's house of personal possessions and telling artifacts, and placing them on the street. This then is a weakening of 'mind,' or the idea of an inner private life versus the public physical body. In setting up an argument to be knocked down, Ryle highlights that in our descriptions of wits, character and higher-level performance of those around us we see their external public behavior as signifying special, interior and more private events. This is the dogma of the 'Ghost in the

Machine.' The key thrust of his critique is that mind is not an entity, a cohesive institution or unique form, but rather is the name for that which emerges from its parts. A mind for Ryle (2000 [1949]: 190) is 'to talk of the person's abilities, liabilities and inclinations to do and undergo certain sorts of things, and of the doing and undergoing of these things in the ordinary world.' There is no domain to be unlocked but rather our behavior is the raw data itself of who we are.

Thus if were to carve up the self with whatever tools and implements of our choosing, we will never find a 'mind' but only parts that give rise to the illusion of mind. In a clever allusion to the ghost in the machine, Ryle uses the example of team-spirit. If on watching a game of cricket one were to go looking for 'team-spirit' we will see umpires, scorers, bowlers, people batting and people catching, and we may even see one team playing particularly vigorously and keenly, but we will not find an entity that we can refer to as team-spirit. The same principle for Ryle applies to mind so for discourses of mind *and* body to be improper topics. Another example that Ryle uses is if we were to say we purchased a left-hand glove and a right-hand glove, we would not also say we bought a left-hand glove, a right-hand glove and a pair of gloves. The same point again applies to physical and mental processes, i.e. there are no special mental processes as it is a fiction of language and its categorizations.

On sentiments, categories and descriptive epithets (such as being kind, shrewd, devoted, wily, cruel, happy, paranoid, compassionate or prudent) these descriptions do not define a thing in itself, but these rather are expressions for a cluster of phenomena we associate with that word. Sentiments refers to propensities, sets of abilities and that a person is prone to doing certain things and behaving in certain ways. It is notable that when we think about the character of a person, this invariably involves doing, behaving and having an effect on others. The link between Ryle, the attack on private minds and behavioral monitoring systems becomes clearer when we consider how the latter work. These too are predicated on clustering dispositions and behavior, by which inferences and predictions are made. This means that in reading Ryle, behavioral systems and the types of observations they make, we are far less interested in the 'nature' of human propensities but rather what they are constituted by. Importantly, these constituents are public. This is a point too often missed. If empathic and behavioral systems read what is made public, they actually *are* witnessing sentiment (whether this is being generous, vain or taciturn) as public phenomena. While we may see empathic media as only having limited access to our private interiors and theaters, this is to misconstrue the nature of sentiment and being that is far more public than our selfish selves believe. The failure so far of empathic media to get closer to us is not due to them being incapable of overcoming private/public borders, but the much milder challenge of reading from without and the honing of machinic intentional capacities. Importantly, from the point of view of privacy norms, this is a more surmountable challenge for developers and harnessers of empathic media.

The links between technologies dubbed 'behavioral' and behaviorism itself are under-accounted for. This is possibly because behaviorism receives much bad press for ignoring consciousness and feelings, and being seemingly cold and indifferent to the richness of human experience. While behaviorism does not entirely evade or overcome these criticisms, its connection with logical positivism and the need for public observability is a useful one. This holds that while we might not be able to know mental events (that are privileged), we are able to study stimuli, effects, operations of discrimination, and make predictions about future actions. Behaviorism is founded on the basis that 'what is felt or introspectively observed is not some nonphysical world of consciousness, mind, or mental life but the observer's own body' (Skinner, 1976: 18–19). This is not to reject the existence of feelings and conditions, but to recognize that these are *collateral* products (Skinner's word choice) of genetic and environmental/cultural history. Possibly the most popular criticism and knee-jerk reaction to behaviorism is that it does not account for the full self and the mental world of the subject. It is to point to over-simplification, but as Skinner maintains, as a criticism this is unfair, not least as a behavioral account will seek to understand the context and environmental conditions that have given rise to a given behavior. This is somewhat reminiscent of Latour in that we do not go searching for spectres (of mind in this chapter's current case) but we trace associations and offer close accounts of behavioral effects. Where Latour points to affective capacity as a means of verification of method and approach, behaviorism makes its point by means of the capacity for prediction. As Skinner also points out, critics of behaviorism are quiet on *what* is left out of the discussion. That is, if behaviorism is an over-simplification, what precisely is it that is left out?

In both behavioral data mining terms and in day-to-day life, to be of a certain disposition is not to possess a specific quality but to be liable towards a particular state. This denial of private minds is of consequence beyond idle philosophical musing. Privileged access is predicated on self-knowledge and inaccessibility of others. It involves knowledge of a superior form and it is that which you cannot have of mine, and I cannot have of yours. It is privileged because it is private. This is crystallized in Berkeley's (1988 [1710]: 193) question in the *Three Dialogues* about the possibility of commonality of knowledge and experience. He asks: 'But the same idea which is in my mind, cannot be in yours, or in any other mind. Does it not therefore follow from your principles, that no two can see the same thing?' This for Ryle is a myth and part of the wider narrative of the introspective ghost in the machine. While there is residual knowledge (for example things about ourselves that only we know and remember), empathic machines (as introduced in

Chapter 7) have a much better chance in principle of knowing us than ourselves. While this discourse may begin to appear somewhat dystopian, this is not my intention. Rather my aim is to use Ryle to highlight the importance of public traces because the informational means by which we buy our socks and pants may, when taken in tandem with other items of data, be much more accurate in depicting our true identity than privileged access methods. If the subject really is that which persists and endures, who is to define it? Is the answer intentional and empathic machines whose *modus* is to pay attention to us, or is it ourselves who busy ourselves, work, enjoy leisure and too often seek distraction (admittedly often during work hours)—ironically by means of the very tools that are interested in our public being?

To rebut privileged access and recognize the relationship between networked system and ourselves in non-'interior' terms is to better recognize the significance of public traces, trends, actions and performances. This is because without recourse to a private and authentic inner domain, what is commodified by behavior monitoring systems is more valuable than first recognized. What is public does not play second fiddle to a private domain packed with *real* meanings and sentiments, but what is on show is what we are. To test this, choose an emotion or feeling and offer an account of how you know that you are in this state: can you do it without recourse to describing externalized behavior or actions (even if not noticed by others)? A consequence of this more public view of the self is that observers may be better judges of our long-term motives than ourselves because it is highly likely we will display biases when making judgments about our personality and character. Whether or not the observing person or programmed system has biases of their own is another question for computer scientists and cultural studies, but what is clear is that our public self becomes much more central to who we are if we deny or at least problematize the existence of privileged access.

Inference tickets

This account of public propensities goes for emotions too as they are not acts or states but tendencies to behave a certain way. With Ryle including moods within his account of emotions, we can also see moods in terms of the language of propensities. This means that the notion of moods in themselves is meaningless, but rather they are a collection or cluster of occurrences and dispositions. When we say that someone is happy or sad this does not mean there is a unique feeling or tone but that a person will display a variety of associated behaviors (possibly telling more jokes and being more talkative if happy). We can make comparisons with the weather. While we might say it is gloomy outside, this is shorthand for absence of sunlight, grey skies, a low temperature, mist and occasional rain. Again, a mood is not a thing in itself but a label for a cluster. In technical and informational privacy terms, this clustering occurs as we visit websites and are tracked throughout our web journey. As we move around, engage in conversations, share things, play games, use applications, and buy and sell things, platforms and advertising systems are able to make rules-based decision about what content to serve by means of our public dispositions, contexts and the clustering of users into highly refined interest groups. The *mood of information* (introduced in Chapter 7) becomes a meta-expression for a wide range of dispositions, behaviors and occurrences, and empathic media impute 'purposes' to our dispositions and actions, and 'understand' our predicaments through these and propose next-steps (e.g. the serving of X advertising material to us).

While we may not be keen on our public traces providing access to 'ourselves,' if we agree that interiority is an untenable premise then the difference between being right and not being wrong may not be entirely distinct (to again appropriate Bateson, 2000 [1972]). Ryle's take on this is stronger as he disallows a halfway house between right and wrong. Here a clock may strike the hour that it is, or an hour that it is not, but it cannot strike an hour that might be correct, but is neither the correct nor incorrect one. Our dispositions, public statements and traces are what Ryle (2000 [1949]: 119) designates (in a non-technological context) 'inference tickets,' which provide license to predict and retrodict. A key point to be noted then is that knowledge about ourselves is *inductive*. To speak of accessing interior selves is mistaken as it gives rise to the idea that the traces and inferred emotions matter less than the real thing. For Ryle, these behavioral traces *are* the real thing, at least if prediction of our subsequent actions is a benchmark of knowledge. While Ryle's attack on 'mind' may be too vehement for many tastes, some reduction of the private subject is required so for the value of what is public to come into focus (otherwise what is considered private hides what is public). However, a fair criticism of the emphasis on the public self is that we all-too-easily lie, perform with great awareness and have some control over our outward representations. This said, our online lives, taken as a net history (pun not intended), would be highly difficult to maintain if it were not a fair representation of our interests and sentiments.

Born public

On the premise of interior monologue, Ryle remarks that the trick of talking to oneself in silence is an achievement not acquired quickly or without effort. It is done by first learning to speak publically, aloud, heard and understood by other people. It is only after this that we might think in silence. Thinking in silence, while often prudent, is not primary or strictly necessary. His point is that the capacity to secure secrecy for what was once spoken aloud in public does not require an interior domain for this to take place, or a ghost in the machine, but rather it is simply not public. Privacy then becomes a convenience, rather than a space or domain. Similarly, while we may be impressed with the intellectual prowess of another person, what they can hold 'in their minds' and their capacity for reason, what we witness is somewhat fallacious or at least miscategorized. Rather, what we see is again learned disposition and the application of logical rules without thinking about them. This is different from taking recourse to a shadowy but impressive place designated as 'mind.' (The same goes for Ryle with subject-specific modifications for boxers, surgeons and salespeople.) Ryle is clear and somewhat scathing in this regard highlighting that 'mind,' is:

> ... not the name of another person, working or frolicking behind an impenetrable screen; it is not the name of another place where work is done or games are played; and it is not the name of another tool with which work is done, or another appliance with which games are played. (2000 [1949]: 50)

Instead the task for both individuals considering themselves, or when watching others, is to understand and ask questions about powers and propensities of actions. Thus it is acceptable to go 'beyond' what is immediately before us (the action) but it is not acceptable in Rylean terms to ask what is 'behind' it. People then are simpler than supposed. Ryle later remarks that if the doctrine of the ghost in the machine were true, 'not only would people be absolute mysteries to one another, they would also be absolutely intractable. In fact they are relatively tractable and relatively easy to understand' (ibid: 110). This is belied by our habits, preferences, gestures, reactions, interjections and, in a face-to-face setting, the various modes of facial and bodily expressions that are public and discernable by other actants.

We do not have to accept all of Ryle's arguments to employ his emphasis on publicness and moods as a cluster or assemblage of dispositions. Indeed, most specialist commentators on behaviorism and mind/body questions agree that it went too far in its restrictions and methodological denial of privacy (Churchland, 1990 [1984]). This is less an admission that mentalist approaches are correct but that the denial of introspection is disingenuous because however rationally confused the premise of introspection may be, it does play a role in life and conduct (particularly in privacy matters and day-to-day interactions with others). Indeed even Skinner (1976), on reflecting on decades of behaviorist research, concedes that internal phenomena do play a role but with the caveats that: introspection refers to the states of our own body and nervous system (not a non-physical reality); introspection only grants access to a small portion of internal states; and that the states discerned are correlated but not causes of behavior. Moreover, while there may not be an inner domain as such—people lie, contrive and create great theoretical edifices, albeit from public languages—people also clearly do experience pain and if my partner has a toothache there is something more occurring than wincing, complaining and displaying a propensity to rifle through a kitchen drawer to look for aspirin. Pain among other states has a qualitative dimension and while there are lessons to be employed from behaviorism for our public empathic account, we do not have to bring on board its extremes. Bertrand Russell (1958: 5–6) also offers key disagreements with Ryle, but first summarizes and agrees that 'the adjective "mental" is not applicable to any kind of stuff, but only to certain organizations and dispositions illustrated by patterns composed of elements which it is not significant to call "mental".' Russell takes Ryle to task on a number of points: for not paying enough attention to the role of introspection and imagination, and for not articulating where thoughts occur if not in our heads. While Ryle talks of externalization and the denial of mind as theater, he also seems unwilling to entertain that premise that the mind is actually the brain. For Russell, Ryle has fallen prey to the pervasive influence of Cartesian dual substance views.

Language-based conceptions of shared experience

We have considered behavior in some depth, but what of the relationship between language, thought and being public? Wittgenstein is deeply influential and discusses the implausibility of private languages and the impossibility of conceiving of mental states isolated from a person's environment. Indeed, a large portion of the philosophical journal-based literature on privacy involves Wittgensteinian questions about language, sensation, knowledge and the public life of meaning. For Wittgenstein thought is deeply connected to language and language is inherently social. In *Tractatus* for example Wittgenstein (2007 [1921] §5.633: 78) rejects the mentalist premise of a subject that entertains and holds ideas. Indeed, he rejects the subject altogether asking 'Where in the world is the metaphysical subject to be found?' caustically following this with 'The self of solipsism shrinks to a point without extension, and there remains the reality co-ordinated with it' (ibid §5.64: 79). For Wittgenstein (2009 [1953]) there is no private theater of sensation isolated from the world and instead what we designate as private requires public criteria. For Wittgenstein, a private language (if conceivable) must not be capable of being translated as it is predicated on interior experiences inaccessible to others. This means it cannot be learnt or translated, but it must make sense to the speaker or knower, and not be explainable in other terms (for example, 'it's a bit like the burning sensation of ice'). By this he argues that language is necessarily social and is something that happens between language users. As a consequence of this it becomes problematic to think of meaning outside of other users, and therefore there can be no private meaning. Conversely, if we do find something to be private, we are unable to put it into language. The rejection of private language has two tacks: the first is the acknowledgement that we might incorrectly remember a novel sensation. So, if we are trying to keep a log of the occurrence of the sensation, marking it as S each time, can we rely on ourselves to recognize and remember the sensation correctly and accurately so to understand that it is precisely S which reoccurs? How do we know it was *that* which occurred, rather than another *that*? Second, what are the criteria for judging S if it is a private language? Further, what does the notation of S tells us about the meaning or significance of S? The notation of S tells us something of the regularity of a sensation needed worth recording, but nothing of its meaning or significance (beyond being notable). It is worth invoking Wittgenstein's (2009 [1953] §293: 106) beetle analogy at this point. Here we have a society of sorts with each member possessing a box containing a 'beetle.' No one is able to look into anyone else's box, and each member says they know what a beetle is only by looking at their beetle. The point is that for Wittgenstein it does not matter what the sensation, pain or beetle might be in the box (or the head), as all we can discuss is what is available to the public. In general, Wittgenstein (along with Ryle) helps to bring about recognition of how social we really are, and the utter dependence of any premise of the subject on public language. As a consequence, the sovereign/private self is somewhat reduced. This is less about pointing to a crisis in liberal humanism (displacement of agency, autonomy, virtue or sanctity of life) or an attack on the subject, but rather the need for emphasis on public life.

Being social

We might also point to the ways in which our own sense of self comes about because of public interaction with others. This places the idea of the sovereign self in some difficulty. My self then is less about 'me,' but also those around me, and how we relate to one-another. Damasio (2011) designates this as 'the social me' seeing the public self as a relational entity. In this sense we are able to see ourselves as living outside of ourselves. Thus while much liberal discourse on privacy privileges autonomy and the individual (assuming no harm to others), understanding ourselves as relational and dynamic entities means greater emphasis needs to be placed on the social environment and the system in which we exist. In regard to interactionism and privacy, this means thinking less in terms of 'my privacy' as a commodity, reserve or stock, but instead caring more greatly about links, associations and relationships. While liberal discourse on privacy privileges the continuous willing self that possesses a strong sense of subjectivity and interiority, we might more properly recognize the ways in which our selves emerge from the context in which we find ourselves. While possibly a fairly simple constructionist observation, it is a correct one that has implications for thinking about privacy. We are thus continuous with the world and this takes us towards an open-system type of view. This moves into the terrain of ecology (or social-ecology as I phrased it in Chapter 8) and that domain of thinking predicated on connection and emergence, and that which identifies the self as part of a much larger network. For Bateson an individual's mind is immanent not only in the body, but also in the pathways and messages outside of the body. In discussion of mind, and what ultimately is an attempt to answer Cartesian dualism, he states:

> ... there is a larger Mind of which the individual mind is only a subsystem. This larger mind is comparable to God and is perhaps what people mean by "God" but it is still immanent in the total interconnected social systems and planetary ecology. (2000 [1972]: 467)

The jargon of co-creation strikes closer to truth than we first thought. The self is not primarily internal or found in neuronal connections, but rather is found in material, social, intentional and linguistic connections. Mind comes into being via external relations. It is a rejection of Aristotelian essentialism in favor of a relational view of how things come to be. This breakdown of distinction between self and the world is fulfilled in that we are always implicated in environments. This open-system or ecological version of the subject might be read at a variety of scales (ranging from the level of particle, wave and attractor to the more familiar ways we relate to other objects and people in the world). The subject in this dynamic view is a construction of its relations with the world and comes to be by means of symbiotic reciprocation. What we designate as individuality and identity is that which emerges out of contact with environments. The idea of a self then is problematic and not easily discernable from the environment with which the person interacts, and the identity of a subject is an abstraction or temporary reification of process.

Conclusion

The aim of this chapter has been to assess privacy in relation to subjectivity and demonstrate that our lives are lived more publically than first imagined. A key way in which we believe ourselves to live privately is by means of private language. This can be understood both in terms of the capacity to inwardly vocalize, but also premises of qualia and unique sensations we connect with experiences. This was problematized by means of Ryle and Wittgenstein, although the excess of their criticisms was tempered by recourse to Russell. Such questions about the possibility of private experience occupy a great deal of the journal-based philosophical literature. The lesson for my account of privacy in relation to tracking, media and technology is three-fold in that: we need to re-evaluate the constitution of the subject; appreciate that the subject lives a more de-centered life than supposed; and therefore recognize that we live more publically than first thought.

This leads me to argue that much of what we consider to be private (e.g. character, preferences and sentiment) is actually quite public. With the subject undermined and the doors of the inner sanctum being very much open, a dualistic or border-based conception of interior and exterior selves becomes less assured. Privacy somewhat paradoxically has less to do with insides and outsides, but rather what we consider to be appropriate involvement of others in our public being. Centrally, as the subject is now somewhat pared back and reduced this has implications for how we value our public traces, and whether being public equates to being open for appropriation by other people and machines. As such, philosophical musing about the ontological status of the subject become political and regulatory questions because a dialectical approach to privacy is predicated on a liberal and sovereign view of the subject. Indeed, our legal and political system in regard to privacy norms is constructed on a border-based view of the subject in which we are first private and then public. With the self displaced from a privileged interior and more public than hitherto thought, what is public is perhaps not as petty or inconsequential as it once was. Do we need to ascribe a stronger set of rights to distributed subjects, or at least ensure proper capacity for control? Should the economic importance of that extended person be more fully recognized in regard to the value they generate? If more of what I am is predicated on expectation and others' readings of these signals, are we able to convincingly reason that our public negotiated self takes on renewed importance in face of a reduced interior subject? In considering the distributed subject we understand that subjectivity is best witnessed by means of patterns of relations and expectations. In an online setting these traces are recorded and compiled by empathic media and used either to generate value in terms of capital, or as the background norm against which unusual traffic may be discerned for state security services. In regard to market-based analysis the question of value is a critical one. It connects to questions about whether we should see value-generation and productivity in terms of labor, and the extent to which we are alienated from the outcomes—if at all. This is the subject of the next chapter.

CHAPTER ELEVEN

Alienation: The value in being public

Critical theory does not respond well to accounts of autonomy as it has difficulty finding an origin or rationale for its existence. Likewise, Marx (2012 [1844]) was somewhat scathing of the idea that rights could be useful in establishing a political community seeing 'human rights' as accentuating an individual's egoistic preoccupations. Liberal citizenship for Marx is to be self-confined to private interests and caprice, and to be separated from the possibility of real community. Liberalism, interpreted in critical Marxist terms, is an inversion of the border-based dyad referred to in Chapter 2 and privileges self-interest over public contributions. This is based on the observation that in its quest to free citizens from public interference liberalism is a philosophy of individualism and separation, and by arguing for a philosophy of withdrawal the liberal project undermines the possibility of emancipation. Marx's argument is that real freedom is found in more positive relations with others, i.e. in community. The relationship between Marxism and privacy thus becomes an uncomfortable one. This has two parts to it in connection with both the social problem that Marxism sees and its solution. On the former, does liberal concern about privacy detract from larger exploitative arrangements in political-economy; on the latter, what value does privacy have when there is no need for rights because the interests of the collective are bound together? Marx (1999 [1867]) is clear in *Capital* that we should not exchange a situation where labor is reified into commodity relations for one where society as an abstract value stands against the individual (where the social becomes more important than its aggregate parts). This means that Marx's aim was a society of free individuals where the interests of individuals and the collective are the same. However, from a privacy point of view, it still remains unclear how this is assured, and whether this form of freedom also grants independence to develop ideas that may even run counter to the collective aim.

Another important point in assessing Marxism in relation to privacy is to factor in alienation. While in Marxism privacy is equated with bourgeoise property ownership (Marcuse, 1955 [1941]), what are the results if we approach privacy by means of alienation? Hegelian and Marxist conceptions of alienation apply well to our informational setting, as well as traditional forms of labor concerns. This is particularly the case as informational industries have, for better and worse, invented novel modes of public communication so to more easily monitor mediated communication and extract extraordinary levels of revenue from the many for the financial benefit of a few. In Marxist credo, alienation involves estrangement from the products we have produced; the act and process of labor becoming impersonal; and alienation from the human race and other human beings. The link between surveillance-based privacy discourses and alienation has commonalities as both involve taking something we have contributed to producing and using this against us. For emphasis, the translators' notes in Marx and Engel's (2011 [1844]) *Economic and Philosophic Manuscripts of 1844* highlight that the word 'alienate' has roots in the sense of a loss of something which remains in existence, of making external to oneself, and to be estranged from. More broadly, such a sense of alienation involves a deep feeling of a lack of control and the sense that the world is not our own, but rather someone else's system. It is that affective sense of violation when things that should be, or used to be, in our control no longer are. Clearly accounts of alienation still have a role to play. With this in mind, this chapter continues by assessing the problematic construction of privacy within Marxist discourse by taking recourse to Hegel. It then inquires into affective, immaterial and autonomist labor perspectives (Hardt and Negri, 2000; Lazzarato, 2004; Berardi, 2009), raising questions about the extent to which value generation equates with labor in relation to media and data mining. The chapter concludes by positing alienation as a more appropriate critical diagnosis of the mining of public sentiment and traces, particularly if we agree with the message of the last chapter that subjectivity and selfhood is more public than first thought, and that the liberal private self has been over-emphasized.

Back to Hegel

Marx was deeply influenced by Hegel who argued that rights are not to be considered as abstract ideas existing in a 'state of nature,' as Locke and Hobbes argued.

Instead Hegel (2001 [1820]) endeavored to provide a foundation to rights by deriving them from a social context. This for Hegel involves property relations, contracts, the economy, family life, the legal system, and broader civic life. The emphasis on social context sees a partial renewal of ancient Greek ideas on the relationship between individuals and the state (i.e. the *polis*), as individual rights (such as privacy) only make sense in relation to wider communities. Although rights make sense in relationships between individuals, these must be relinquished to the state as this represents a meta-reality of sorts to which citizens do not need to appeal for rights. While Hegel does not discuss privacy directly, he is clear that people who belong to the ethical and social fabric of society are to be afforded freedom. Participation and citizenship is key and in *The Philosophy of Right* there is a telling section that highlights Hegel's requirement that free will can only be enjoyed in the context of full engagement with civic life (i.e. networks of property, contracts, business, moral commitments, the legal system, and engagement with the government). Indeed, withdrawal or seclusion is deemed impossible 'because no one can succeed in alienating man from the laws of the world' and that 'we cannot believe that the odour of the world of spirits does not in the end penetrate their seclusion, or that the power of the spirit of the world is too feeble to take possession of even the remotest corner' (2001 [1820]: §153, 137). In Hegelian terms, as in ancient Greece, value is placed on being public and this occurs through engagement with property ownership, and the public economic and legal matrix this embroils a person within. Being public by means of property is thus an expression of private will. Indeed, in *The Philosophy of Right*, Hegel sees free property as a fundamental condition for the state to flourish. Hegel saw property as the only way to express private will, as property is that which makes the will manifest and, for Hegel, 'is rightly described as a private possession' (Hegel, 2001 [1820]: §46, 58). As property is a means of establishing will, property is to be seen as 'this' and hence as 'mine,' and in relation to property 'there is no room for any reference of the will of others to my will' (ibid: §113, 100). Hegel offers sensible caveats, for example the theft of a loaf of bread to prolong life is a very different case (not least that the denial of life is the utter denial of freedom to another). Indeed, on this point, alienation comes about because of the lack of ownership of property.

However, property ownership is responsible for causing a schism with the state that means the state becomes something alien. This is because the state, as an assemblage of private interests, is called upon to defend the rights of property owners. This gives the state separateness and independence *over* its members, and therefore becomes alien. Property thus becomes a problem for freedom, not an answer, because property is that which ties individuals to private interests rather than public interests. Marcuse (1955 [1941]) sees in Hegel (drawing on

Theologische Jugendschriften) a view of property that will come to oppress property owners, along with those others close to them. This is a dead world that gets in the way of enjoying authentic relations with others and as a consequence blocks a proper sense of community. As Marcuse tells, this is the earliest formulation of alienation that will come to play a role in later Hegalian philosophy. Marcuse also discusses Hegel's ideas about the relationship between 'inner freedom' and the need to be part of the state to properly realize humanity where rights are no longer necessary. Within this is a fundamental Marxist argument that liberal rights (based on autonomy and freedom) are mythical and bourgeois, and this hides the fact that only a socialist society can deliver freedom. Marcuse (1955 [1941]: 12), citing Hegel's *Erstes Systemprogram des Deutschen Idealismus* from 1796, highlights Hegel's mechanistic metaphors in which the state would come to 'treat free men as cogs in a machine.' We also see early indications of Marxist notions of alienation in that the state exists as an *estranged* entity as the political interests of its citizens eventually disappear altogether.

> Inner freedom does at least reserve to the individual a sphere of unconditional privacy with which no authority may interfere and morality does place him under some universally valid obligations. But when society turns to totalitarian forms in accordance with the needs of monopolist imperialism the entirety of the person becomes a political object. Even his innermost morality is subjugated to the state and his privacy abolished. (Marcuse, 1955 [1941]: 199)

All too familiar accounts of bureaucracy, commodity-production, calculability and dehumanization as the flip-side of the will-to-rationality come to the fore. Freedom still matters, but it has become that which is provided or allowed by an agency (the state), rather than being a condition in which we live. Again, this is due to property ownership, and that freedom of contract can only be guaranteed by force and threat.

Alienation

Alienation does not allow for the full and free development of the individual. The premise of alienation in Hegel has Kantian roots (along with Fichte and Schelling), due to Kant's (1990 [1781]) Copernican turn in the *Critique of Pure Reason* and the impossibility of subject/object relations in transcendental idealism. Less wordy, this means no discussion of what is real can take place without awareness of the role of the human mind in creating reality and knowledge. Hegel's (1910 [1807]) observation is that we are alienated from own consciousness, and we must go through a philosophical and phenomenological process for consciousness to understand itself (hence the phenomenological interest in introspection and experience). Consciousness in Hegelian terms involves our relationship to ourselves and also to things (objects) in the world. The relationship to Kant is by Hegel's agreement that objects conform to our cognition and the ways in which our faculties create the world. In Hegelian terms we are alienated from ourselves and only by means of a phenomenological journey may we overcome self-alienation, and progress towards what Hegel accounts for in *Phenomenology of Mind* as 'absolute knowing,' characterized therein as a divine reconciliation between minds and objects in the world. Analysis for Hegel involves understanding about how the subject discloses the object, or why it appears to us as it does. The Hegelian method is that mode of introspection that assesses the relationship between the subject and the object, and draws out contradictions from this experience. This means that rather than examining what is simply 'present,' one instead should understand the appearances of the mind to itself and begin the road to absolute knowledge. This is a process of sublation (*Aufheben*), which is the progression and process of abolition that takes place as thesis and antithesis are dialectically played out in history. The absolute is sublation because this, for Hegel, remains ahistorical. Once we understand how these appearances have come to be, we can begin to understand and reason the nature of history (reality) we have created for ourselves.

Marx (1999 [1867]) famously upended Kantian and Hegelian subjectivism (in the introduction to *Capital* he describes his account as directly opposite), focusing on the material dimension of life and its effects on mind and cultural life. Where Hegel focused on mind/spirit where sublation is worked out, Marx saw this resolution in materiality. Marx's critique very much begins from the realm of objects, and his target of historicality is the capitalist nature of the social relations he witnessed where an unfettered market economy dictated these relationships. The relationship between worker consciousness and social existence is held to be false, and it is this falsity that has to be overcome for true and proper relations to come into being. Here we can see Hegel's dialectical influence at work as contradictions are worked out. Marx thus witnessed workers producing greater wealth for others, yet poverty (spiritually as well as economically) for themselves by means of these social relations. Workers thus became less like people, and more like the commodities they worked so hard to produce. The point is perhaps less about levels of wages, but rather the miserable mode of labor itself in the modern society that Marx witnessed.

Alienation as a critical premise also has roots in Feuerbach and Stirner. For Feuerbach (1957 [1841]) alienation involved religion and the ways in which the attribution of human qualities to God meant that people are unable to recognize that the qualities ascribed away are actually qualities that belong to them. Alienation is very much connected with what Marx and Engels (2011 [1844]) account for in terms of de-humanization, instrumental accounts of labor, and a split between subjects and objects. This sees 'armies of laborers' and wars between capitalists. For wars to be won, workers cannot be treated as men but instruments of production required to yield as much value as possible at minimum cost. It is worth being clear too on the situation that classical Marxism refers to. This is labor that is not only invasive or unpleasant, but that which is long, painful, sometimes disgusting and seemingly characterized by being paid less for more taxing work. We might then proceed with a little care and possibly respect before we start using terms such as 'immaterial labor,' particularly in relation to social media use. However, the broader principle is correct in that in order to compete, corporatists pay their both material and immaterial laborers as little as possible while trying to yield maximum return. On estrangement, Marx and Engels comment:

> ... we have shown that the worker sinks to the level of commodity and becomes indeed the most wretched of commodities; that the wretchedness of the worker is in inverse proportion to the power and magnitude of his production; that the necessary result of competition is the accumulation of capital in a few hands, and thus the restoration of monopoly in a more terrible form ... (ibid: 49)

Labor and capitalism also have self-creating and self-generating qualities (more begets even more of the same processes). These processes go further than self-production and transform their environments because capitalism, taken as a social fact, produces more of itself and alters that around it. Historically this has been seen both in the corporeal and digital environment that in a very real sense have both been transformed. This is where links can be made with Heidegger as capitalism seeks to make present-at-hand, as evidenced in the shift from open networks where anonymity was the social norm to intense data mining, and by means of what are conceived in this book as intentional machines and empathic media. In a classic Marxist sense the worker is reproduced as a commodity and this of course happens online too as more commodification begets more commodification (never a reversal). We should also consider what the product is. Offline, labor is said to congeal into a material object of labor that can stand *against* the worker that has produced it. The product becomes alien and the sense of distance or estrangement from what one has produced is alienation. What then is the immaterial data-based product? I propose two candidates: 1) advertisements; 2) more processes of commodification. This is quite different from what Marx and Engels conceived, but these are very different times. Evidence of us being-public is to be found in advertisements that are subsequently placed before us with traces and congealed elements being both external to us, yet employed to persuade us to part with real wages. In effect, we are selling ourselves to ourselves. If we accept this as an update to the Marxist thesis of alienation, then we might marry this with Marx and Engel's despising of marketing. On this they remark that a producer 'puts himself at the service of the other's most depraved fancies, plays the pimp between him and his need, excites him in morbid appetites, lies in wait for each of his weaknesses—all so that he can then demand the cash for this service of love' (ibid: 82–83). The mode of seduction as well as production was clearly alive and well in the mid-parts of the nineteenth century, and pre empting the twentieth so often linked with innovation in manipulating desire.

Such enticements were seen by Marx and Engels as an incremental stripping of dignity and baiting of one's self away from one's own being. They would then be aghast at the state of the contemporary commercial communications industry that uses public dispositions, preferences and honeypot aspirations inferred from mediated communication and mobility, with which to sell ourselves back to ourselves. However, while horrified, Marx and Engels might not be surprised, as the current milieu readily squares with their theories on alienation. To quote once more:

> The alienation of the worker in his product means not that his labor becomes an object, an external existence, but that it exists outside him, independently, as something alien to him, and that it becomes a power on its own confronting him… (ibid: 51)

Alienation does not just involve objects and outcomes, but also processes and the act of production itself. Indeed, this is the more dehumanizing aspect of the thesis. Here the mode of production makes one's self a stranger by turning the individual into something else (physically demanding 16 hour days in the nineteenth century took their toll). The point on self-perpetuation is also noteworthy as the 'worker produces capital, capital produces him—hence he produced himself, and man as *worker*, as a *commodity*, is the product of the entire cycle' (ibid: 60 [emphasis in original]). This intensity and form of production not only alienates one's self from self, but involves a sacrifice of physical and spiritual life, and what it is to lead a productive life (characterized by a greater degree of freedom and not just to exist). Clearly this sits at odds with early twenty-first century notions of immaterial labor that do not equate to drudgery, nor are people forced to use social media and related platforms, although quite arguably they do feel compelled. There are other differences too and where wages were once tantamount to oil used to service and maintain a machine, digital industries see zero financial payment and no need for 'wages.' Indeed, what is even more impressive in contemporary capitalist processes of value generation is that we pay for our smartphones, tablets and other technology that engender more data, value and thereby capital.

Abstract alienation

The success of capitalism derives from the way in which the real world of things and people are exchanged into quantitative representations. This process of alienation-as-reification connects Marx (1970 [1859]), Lukács (1967 [1923]) and Heidegger (2013 [1941–2]), as each critiques the transforming of aspects of the self into that which is present-at-hand, objective and subject to manipulation. It is to exchange the idea of value into exchangeable quantity, or more plainly quality into quantity. For Fromm (2002 [1956]) alienation is a fundamental feature of capitalism. He quotes Marx in *Capital* on alienation remarking that this is a condition where a person's 'own becomes to him an alien power, standing over and against him, instead of being ruled by him' (2002 [1956]: 118). While Hegalian and Marxist-inspired, there is a slight difference in emphasis as for Fromm it has less to do with the results of one's handiwork, but rather the experience of the self as an abstract entity. Here alienation is the way that we experience ourselves as an alien. In *The Sane Society* Fromm focuses on the ways by which we do not experience ourselves as the center of the world; how we are estranged from ourselves; and consequences of being out of touch with ourselves.

Adopting a Marxist viewpoint on immaterial labor and informational industries (Hardt and Negri, 2000; Lazzarato, 2004), one might argue that little has changed. Although one might question the nature of labor, if we can agree that labor involves externalizing one's efforts into a service or product of some sort that expropriates and generates value for another person, then capitalism is functioning as it always has done. Indeed, the more that laborers produce, the greater capitalism becomes and the less powerful its victims or subjects become. While one might question whether Facebook users are really victims, what is clear is that our labor congeals into products that are alien to ourselves. Indeed, this is the promise of data miners as they tell us that our information has been aggregated into that which is unrecognizable to us. This is partially true in regard to the traceability of the information back to us, but false in regard to the marketing and advertising that represents our public selves, behavior and wishes that are aggregated and assembled, and cast back at us. Thus our actions (disclosing, posting, choosing, clicking, pointing, swiping, enlarging and so on) are converted into useable form by data miners and aggregated into informational objects that are displayed to us (advertisements/marketing offers). The consequence of this is that a person sees aspects of himself or herself in informational objects that are produced in part by the monitoring of mediated and externalized subjectivity. This is a long-winded way of referring to that mildly spooked feeling we have when behavioral advertising is too close in sentiment for comfort. The unpacking is important though as it highlights the external nature of lived subjectivity, the appropriation of the external and distributed subject, the reification of subjectivity in the object, the independence of the object that is imbued with our own subjectivity, and the process by which we come to stand against ourselves. We thus become alien to ourselves. Moreover, the more we labor, the more powerful and discerning the alien world of informational objects and procedures therein become.

Marcuse (1955 [1941]: 277) discussing *Capital* highlights that 'the fact that it [labor] becomes the property of another bespeaks an expropriation that touches the very essence of man.' What are we to make of 'the very essence' if we take seriously the message from the last chapter that we are more public than first thought and our being comes to be by means of connections and systems of relations? What are we to make of those empathic industries predicated on understanding sentiment, moods and exteriorized being? In evaluating the ascendency of data mining systems and computer networks, Stiegler (2010) builds his arguments around retentional systems and grammatization. Retentional systems are those that record, and grammatization refers to the ways in which we exteriorize memory and aspects of mind whether this be on stone, papyrus or silicon in computer networks. Adopting a very Heideggerian tone, the focus of criticism here for Stiegler (2010) is the 'technicization of memory' and the ways in which social and [bio]political arrangements seek to control and harness 'the energy of the *proletarianized consumer*—that is, the consumer's *libidinal* energy, the exploitation of which changes the libidinal *economy* and, with it, the economy *as a whole*' (2010: 25 [emphasis in original]). In political economy, these arrangements for Stiegler become something more than the more usual designation of 'businesses' or 'regulators,' but rather they become psychic organizations. Psychic labor or what might otherwise be discussed in terms of immateriality, cognitive, affective and creative capitalism is reduced to calculability, and structural algorithmic logic. For Stiegler, '*logos* has become, pharmacologically and economically, *ratio*' (2010: 46). However he does not just lament this situation but suggests a move towards 'a new conception of economic value, and of its *measurement*... that it is not reducible to calculation' (ibid: 51). This appears to me like a step in the right direction, although I am less sure about the libidinal phraseology that invokes Lyotard's (2004 [1974]) *Libidinal Economy* and psychoanalysis. My agreement is more mundane on the recognition that public behavior traces are valuable because, quite arguably, that is what we as subjects are—a public assemblage of traces. I argue then that if we are to take lessons from Hegelian and Marxist-inspired theory, consciousness of alienatory processes in this setting involves awareness of the value of public traces because, with the subject reduced, this is what we are. As such, with the bubble of the liberal subject popped and reduced, greater care for what is public is required.

Criticisms

Discussion of labor versus immaterial labor is uncomfortable because the two types of labor are quite different, with one of these potentially involving being horizontal on a sofa. Indeed, I like Fuller and Goffey's (2012: 68) phrasing of this as an 'epidemic of productive fidgeting that leaves logs, cookies, click trails, and identities in its wake.' However, the principle of labor relations is the same in that things laborers create fill our social world and generates economic value for others. This world created of our public selves serves the economic interest of others, despite the fact that it is made up of aspects of our own being. Our actions and labor then are objectified into a world that does not act in our interests. However, the question of whether this is labor is very real. While Google searching, Facebook posting and tweeting creates value, are we to accept that labor is defined in terms of that which generates value for another? While clearly these processes are productive, are they *really* work? This seems to me a weak and spurious reading of Marx, and somewhat inconsiderate to people in poor working conditions today. I argue that alienation is a more accurate and useful focus for critical attention than immaterial labor. If critical emphasis is shifted to alienation, we are able to take the best of Marxist critical theory as a means of investigating mediated value generation, expropriation and reification—and those processes by which personal and social relations become congealed and used as a means of selling ourselves to ourselves.

Conclusion

This chapter has explored and unpacked the notion of alienation in relation to Marxist critical political economy. It dodged early on the problem of privacy in a Marxist vision of community by foregrounding the efficacy of alienation as a critical premise with which to consider privacy in regard to the self who is more public than first thought. At heart of alienation is the relationship between people and things in the world. In a Hegelian neo-Kantian sense, the subject played a key role in constructing the objects. Hegel's recommendation was that we consider more carefully the means by which we constitute and create objects. He asked us to do this by means of questioning our consciousness and experience of the world. Marx upended the subject/object relationship requiring that we better understand how material objects constitute subjects and their consciousness's. Both Hegelian and Marxist forms of alienation involve recognition of a split and having worked through these approaches I argue that we should shift critical focus from ideas of labor in informational environments to more intense scrutiny of alienatory processes. This becomes imperative if we agree that the sovereign subject is somewhat reduced, and that greater care is needed for our public selves.

CHAPTER TWELVE

Spinoza: Politics of affect

As argued throughout, all accounts of privacy should take into account its affective dimensions. This means we should not only inquire upon ideas about ethics, questions of autonomy, political-economy, emergent protocol, processes of making-transparent, co-authorship, the ontology of data-based knowledge, and questions about the location of the subject, but also the existence of experiences of privacy. Privacy does not only involve the protocols of a system or arrangement, but also the tangible dimensions that go along with these when there is breach or discord. Where earlier chapters have assessed ethics and epistemology, this chapter addresses the role of *feelings* in relation to privacy arguing that the material and affective dimension of privacy should not be omitted. If skipped, we would be left with a very blank presentation of privacy. This would be a de-charged account leaving it prey to a one-dimensional discussion of systems, norms, processes, and the somewhat cybernetic language of tacit and overt optimization of protocol between actants (none of which are incorrect).

While behaviorists may deny the premise of mind and mentalism, when a privacy event occurs it certainly feels very personal. It has phenomenal attributes in that, at the very least, it appears to us an acutely subjective experience. Indeed, in undergoing a privacy event in public we may breathe deeply, quell heart rate, cool blood vessels and relax our face muscles for no one but ourselves to know what has just occurred. Exploration of visceral and affective undergoing finds expression in a range of privacy and surveillance scholars, particularly in regard to those of a Foucaultian persuasion. This continues to be the preeminent approach for surveillance studies (Elmer, 2012), even surpassing Deleuze's (1992) re-evaluation of Foucaultian (1977) societies of control and dispositifs—although both in their own ways are concerned with understanding power, biopolitics, coercion, regulation of the self, enforcing of conduct and encouragement of docility. While acknowledging Foucaultian influence on affective premises of surveillance and privacy this chapter takes a different tack, instead assessing the influence of Spinoza. Work on Foucault is too well-known to be rehashed here, but for those unfamiliar the journal *Surveillance and Society* has a comprehensive special issue on the Foucaultian influence on surveillance studies (see Wood, 2003). Instead, this chapter traces affective experience in relation to the philosophical construction of affect—particularly in regard to Spinoza and Deleuze, especially Deleuze's (1988 [1970], 2011 [1969], 2011 [1981]) work on affective bodies rather than his aforementioned essay *Postscript on the Societies of Control* from 1992. This is so to more fully recognize and unpack the quasi-material dimension of surveillance and also clarify 'affect,' a much-abused term.

Affect involves a refusal of abstraction. It is that which will not allow us to depart, deny, lose sight of or reject the lucid sense of a privacy event. This is important, as it is too easy to take theoretical recourse to accounts of context, systems, tracing information flow and that which abstracts from the lived dimension of privacy. Clearly affect is a messy and imprecise topic, but it should not be ignored for the sake of precise modeling. Instead I want to meditate on the emotionality of a privacy event, how this comes to be, the situation or event it arises out of, and use this reflection to guide understanding. Thus, in relation to privacy and affect, it is important in any understanding of privacy that we do not lose sense of the tangible dimensions of privacy that are defined here in terms of *phenomenal materialism*—or that lived sense of the material world in which we reintroduce the 'being' part of being private back into understanding. This is to better understand the affective poetics of privacy and avoid overdependence on representational modeling (diagrammatics) of contexts, social arrangements and information therein. Privacy might not involve being touched, harmed or physically intruded upon, yet still generate a visceral reaction—sometimes best characterized by high levels of distress and other times just feeling a little creeped-out.

The influence of Spinoza

This visceral and corporeal aspect of privacy and protocol breaches are readily theorized with recourse to Spinoza. Spinoza (1996 [1677]) is a determinist who believed that the universe is a whole in which every action and event is the necessary outcome of the actions that preceded it. While earlier in his career he subscribed to Cartesian substance dualism, he came to argue for reality as a single substance in the sense that there is a commonality that underlies all existence (be this of God or more familiar things such as trees or carpets). Indeed God for Spinoza *is* the natural world and his God is not about rule or providence but rather that which is beginning, end and the deterministic system itself. For Spinoza physical or mental 'worlds' are extensions of one and the same substance, and all are equally subject to causation. What passes for free will in Spinoza is awareness of our appetites by means of a reflective capacity ('If I do X there is Y consequence; but if I choose A there is B consequence'), but for Spinoza choice is illusory. He has enjoyed renewed popularity because of the means by which he bypasses questions of dualism, and his neutral monism (stuff that is neither physical or mental) and ideas about affect. This involves Spinoza's sidestepping (or overcoming) of materialism and dualism, and the bringing together of all of this under one common banner. In regard to how he has been adopted today by philosophers and cultural theorists his determinism is rarely mentioned (determinism is another uncomfortable topic for cultural studies and critical theory), but his suggestions about bridging mental and physical phenomena have appeal.

His ideas about bodies have been very influential both on Deleuze and the vast gamut of Deleuzian-inspired literature. Indeed, some of Deleuze's more difficult writing on vectors and immanence finds initial expression in Spinoza. In *Ethics* Spinoza (1996 [1677]) argues that stones, animals, humans and God all share the same substance and there is no plurality. What we call bodies are regions of extension that are closely related when we move, rest, turn left or go right. These are also *paralleled* by fairly cohesive sets of ideas, and when this concurring ensemble causes something else we think of them as single things. Such an argument has been an opportunity for later thinkers who seek to span mental life and materialism, and develop a language of affect by which the separation of mental life and materiality is transcended. Affect then applies equally to what we routinely call mind and body, and when these are both engaged. Deleuze's (1988 [1970]) monograph on Spinoza offers an excellent account of both Spinoza, and Deleuze's own work on the plane of immanence. This is a plane characterized by bodies, but as Deleuze develops here and elsewhere this is not a familiar account of bodies (with wings, arms, tentacles and so on). Rather a body comes to be by its capacity to affect and be affected (and clearly there are links to be made with Latour). Further, following Spinoza (1996 [1677]) himself, these bodies themselves are mosaics (akin to Latour's black box) and thus an assemblage of seemingly simpler affective bodies. To say that things are made of things is not an especially astounding philosophical statement, but to account for a given body as made up of interchanges of affect and for a body to subsequently be part of wider affective processes is to say something more. It is to move from things not just to relations, but also to a more charged and inter-dependent view of life. When we start to think of bodies as affective assemblages or as temporary unities (be this for a seemingly long or short period) characterized by internal jostling, interplay of bodies that comprise the meta-body and disruptions therein, what passes for a body (or actant) is widened. This has consequences for discussion of the subject and agency (also see Bennett, 2010). By adopting a more granular approach to bodies, we might take ourselves as the example and see that what passes for will and agency is effected by means of the interplay of a wide range of bodies inside our skin and outside of it, whose composition depends on proteins as well as pronouns. Will and agency then are demonstrably a human/non-human assemblage and certainly we as individuals are not sovereign or pure. Individuality thus becomes more porous or, indeed, public as we understand the self to be a confederacy of multi-scalar processes.

As a point of philosophical orientation, Spinoza's line of inquiry assumes a mind/body parallelism, and it is this that allows Deleuze to build his own arguments on the value of immanence. This grants us an opportunity to see privacy in far more visceral terms because we are much less constrained by substance dualism (representations versus materiality), and this opens up opportunity for ontologies of tactility, intensity, sensation and affect. This is because privacy events involve the breaching of protocol, and when experienced are intensive and sensational situations that noticeably involve both cognition and pre-cognitive bodily reactions. Indeed, they may encompass visceral reactions that are pre-conscious and pre-reflective. Privacy experiences and events thus become more potent *because* of absence of defined and easily delineated meaning. This point is dramatically made in the German 2006 film *The Lives of Others*, directed by Florian Henckel von Donnersmarck, set in East Berlin as the main protagonist, a playwright named Georg Dreyman who is convinced he is not under surveillance by the Stasi, finds out after the fall of the Berlin Wall that he was surveilled for years. His reaction to being told that his home was always bugged and under surveillance highlights that it is the autonomic aspect of reactions to privacy events that makes 'it' difficult to deal with. Hence when we engage in empirical work on privacy and ask 'what do you think privacy is' this is the wrong question. Rather the first and most important question is 'what do you feel in regard to the behavior of X....' By shifting away from representational approaches to privacy, we move towards that parallel Spinozan approach that assesses interaction between body and mind that in more contemporary terms involves better understanding of behavioral, chemical and neural responses to stimuli, situations and contexts, as well as insights generated from self-reports of phenomenal experience.

Accordingly our reactions to situations we describe as involving privacy are always immanent and affective because, however slight, there is an accompanying emotion and feeling as we square the situation with our map of how contexts and protocols should be. This is also somewhat Levinasian (1999 [1974]) as his phenomenology (account of experience) provides greater materiality, tactility and sensibility that comes to be through enactment and proximity with others. The key point here is that way in which privacy protocol overflows representation so to be more than cognition or memory of how to behave, but is sensual and affective. Reactions to privacy events may not be easy to articulate and are possibly best elucidated by poets and artists than academics, theoreticians and lawyers, but they certainly involve sensation, stimulation, disruption and intensified moments. They are also pure in that when we experience a given event they are unqualified and immediate. Privacy is intensive in that it is a mode of experience that in affective terms cannot be indexed to anything outside of itself. While we may explain privacy in terms of property, reputation, control, access, borders, context, liberty, historical norms, ethnocentricity and so on, the affective dimension of privacy transcends each of these. This type of affect possesses a purity that while difficult to phrase (although *affective breach* is my working definition), does have a discernable quality.

Privacy then moves beyond the domain of the abstract and intellectual so to be involved in that which is immanent and corporeal. It is that which has occurred before we have processed the event and made sense of the situation and protocol breach. It transcends mind/body dualisms and joins those affective states such as anger, happiness and revulsion that eschew misleading dyads. Privacy as both an immediate and immanent phenomenon emerges as a reaction to events and interactions with other subjects and objects in the world. This is where Spinoza (1996 [1677]) is so valuable and worthy of interest because some time he ago argued that mind and body are not autonomous but univocally work in parallel. This leads to the correct view that mind is an inseparable part of the organism itself. In contrast to dual substance views (mental and material) that are predicated on the lack of extension of the mind and material matter that cannot think, a substance monist position combines the two. Making the argument for this is easier than accepting it as so much of our language, law, religion and common-sense thinking is predicated on a mind–body split. This monism has consequences because it raises the status of the things, objects and representations we meet every day. Be this the Google homepage, the building my office at work is in, the people I converse with, or the words they utter—these can all be seen as *bodies* capable of affect. This is to invoke the Spinozan and Deleuzian (1988 [1970]) notion of the plane of immanence, and provide all we encounter with a quasi-materiality that comes to be because of the capacity for affect. More simply phrased: if it is capable of affect, it is a body. Affect thus has to do with the ways in which we move or are moved and we need not be in proximate range, but rather ideas, words, information, images and symbols from a distance may also affect.

A privacy event is an affective encounter that generates a shock and unmapped situation. The event is an intensive aberration of systemic protocol and thus ontologically spans the old divide of materialism and metaphysics. Such events are commonplace, possibly involving finding out that information has been made public that should have not, or when our skin and bones selves are exposed at an inconvenient moment. Affective events have corporeal properties demonstrated by the fact they may consequently affect other bodies. One might avoid speaking openly with a colleague because they have disclosed impertinent information about another. Or the leaking of electronic information may lead to the meeting of skin-and-bone bodies to decide new security protocols, alter software, restrict access to hardware, and possibly even the drafting of new legislation that has consequences for all sorts of bodies including people, businesses and organizations. The disruptive event itself takes on affective bodily form (to reiterate—an outcome defined by capacity for affect) that tangibly influences, alters and changes that which is around it. As co-emergent phenomena (involving things + people + processes + us), privacy can be seen as a *bifurcatory event* that transforms selves and environments.

Expectation

Affective conceptions of privacy also have much to do with management of expectation. This is an ontological critique in that it deals in what we perceive either to exist or not exist for us. Indirectly Rössler (2005) provides a warming example of this by means of depicting a boy scout who perceives himself to be alone in a room, but is instead a subject for a gazing voyeur. The scout in the room is in the center of the room executing military commands exclaiming 'About face!' 'Right dress!' 'Ten-shun!' 'Parade set!' He then proceeds to cavort around the room, assuming that nobody knows or sees his behavior. While a silly and harmless example, it illustrates well the need for respite from observation of others to behave freely. Thus while autonomy-based accounts tend to be framed around us being productive and writing some masterful explosive political thesis, exploring deviant and exciting behavior, corporate exploitation, and shielding from data processing by state-initiated global surveillance systems, privacy and autonomy can also be thought of far less taxingly in terms of respite and the freedom to dance around the kitchen to uncool pop songs. It highlights the ways in which radical transparency is an ugly proposition not because what one does not wish to share is immoral, but because we would simply feel silly. While not a startling philosophical insight, the argument and observation is correct and to be in a state of 'always-on,' is to never be able to truly relax. The presence of others, whether this be actual, mediated, remote or even imagined will affect, alter and modify our behavior, and it is quite reasonable to assert that as an ongoing situation this is not a desirable way to live. Sartre (2003 [1943]) explores this directly in *Being and Nothingness* remarking on the consequences of 'the look' and the ways in which such gazing objectifies, pins us down and reduces us. In surveillance and privacy terms, the look is also that which can curtail behavior, authentic expression and liberty. Sartre remarks that this alienates us from ourselves forcing us into what we might refer to as *third-person being*. Indeed, on this point we unwittingly invite comparisons with panoptic metaphors. In this we develop an extra dimension of consciousness, or a reflexivity, in which we more carefully surveille ourselves in reference to other possible surveillers. Third-person being thus curtails all sorts of things. It is well-known that students clean up their social media profiles before leaving university to seek employment. Beyond compromising photos of nights out, this also manifests in not appearing too political or anti-capitalist, with some even expressing concern about attending real-space demonstrations for fear of being photographed and tagged (Turkle, 2011). While we might attribute only a little significance to this, it is worth a second thought if we bear in mind that liberal ideas of privacy as developed in Chapters 3 and 4 are predicated on having space to develop ideas and novel beliefs away from influence of the majority view. Each of these examples connects to *affective privacy*, as systemic protocol becomes not just an abstract set of codes or preferences, but quasi-material guiding norms. Protocol while insubstantial possesses affective properties and is an actant or black box in its own right.

Watching lives of others

Affective privacy does not just involve the watched but also the watchers. In the aforementioned *The Lives of Others* this point was made in relation to the other key protagonist, Stasi Captain Gerd Wiesler, who broke Stasi orders to protect the writer and his partner he was surveilling because he witnessed and sought to cherish moments of beauty while surveilling. Beyond the German Democratic Republic, we can again cite social media usage. Turkle's (2011) interviewing of teenagers also raises a number of privacy events, particularly in terms of surveillance, stalking and affective privacy. In Facebook usage 'stalking is a transgression that does not transgress' in that it has become normalized, but remains somewhat creepy (Turkle, 2011: 252). In her study of adolescent and adult social media use, one of Turkle's interviewees remarks that while it is not against Facebook rules to look at other people's wall-to-wall conversations, and most Facebook users do it, 'it's like listening to a conversation that you are not in, and after stalking I feel like I need to take a shower' (ibid). Stalking, or less sensationally the act of watching and listening in to interactions that one has not been invited to, has become a new voyeuristic norm. What is most interesting and instructive about this example is that the affective dimension of the privacy event does not have to occur for the actant that is being watched or accessed (I use actant as it could be an object that is being looked at, for example a computer file). Instead those who undergo affective involvement are those who instigate the breach in protocol.

Voyeurism is not blank analytical watching but an emotionally charged situation. Where critics of privacy ask 'what have you got to hide' we might ask the surveillers (which will include most of us at one time and another) 'why do you want to watch' and 'how do you feel when you do this'? Sartre develops this point with granular flair as he articulates the experience of the watcher immersed in an affective act encompassing both mind and body. Positing himself as the hypothetical watcher, Sartre (2003 [1943]: 283) articulates that 'I am my acts,' and that he is 'caught up in the circuit of my selfness' as 'consciousness sticks to my acts,' and that he is imbricated in 'a pure mode of losing myself in the world, of causing myself to be drunk in by things as ink is by a blotter.' By being immersed in the act, the self is reduced to a nothingness ensnared in the objectified situation. For Sartre's voyeur this emptiness is brought into focus and then lost by what comes next—footsteps! (Or to expand the example, recognition by a third party.) All sorts of parallel mind/body modifications occur as the self sees itself from outside itself, or as someone else would see that self. And then shame ensues, or that understanding with which the subject realizes they are an object for an Other. The subject is caught, trapped and alienated from the domain in which they are present as the Other looks on at the detritus of surveillant behavior. As Sartre puts it, this involves 'internal hemorrhage' and 'flow of my world toward the Other-as-object' (ibid: 285). As the watcher's watching is made public, a unique affective situation comes to pass generated by co-presence in which the subject is offered up for appraisal and judgment by the third party Other. The subject in this situation undergoes a dramatic loss of freedom, that only moments before was experienced as effortless nothingness. The subject is fettered, exposed, in danger and now being-for-another. And who is the Other? For Sartre, it is 'alienation of my possibles' and 'if I am wholly engulfed in my shame, the Other is the immense, invisible presence which supports this shame and embraces it on every side' (ibid: 293). We might add that the Other of course does not have to properly exist and the creak of a door may be a false alarm and a virtual Other, but the affective principle or the *surveiller affect* is the same: the Other may be omnipresent, and the surveiller continues to experience being-for-others. This is the meaning of affect in the surveiller affect—it is the experiencing of locked-up selfhood and existing for a real or imagined Other.

Conclusion

While the usual point of departure for affective accounts in relation to privacy tends to involve Foucaultian (1977) biopolitics, this chapter followed another route going back to Spinoza (1996 [1677]) and subsequently Deleuze's (1988 [1970], 2011 [1969], 2011 [1981]) work on bodies. This emphasis was utilized to ensure that parallelism and the physical as well as mental aspects of privacy experience are properly recognized. The ontological shift shares commonalities with discussion of flat ontologies defined in terms of the capacity to affect in Chapter 8, and the connection between Deleuzian bodies and Latourian actants is worthy of note in this regard. Seen this way a privacy event is an occasion that spans the old divide of materialism and metaphysics, and for the self generates a shock and unmapped situation within the nervous and cerebral system, yet may also impact on a diverse range of multi-scalar bodies. This then is to depict privacy in terms of immanence and raise the status of the affective dimension of privacy so to foreground its political importance in decision-making about privacy protocol governing any given system.

CHAPTER THIRTEEN

Whitehead: Privacy events

This chapter develops the logic that underpins affective, systemic, contextual, assemblage-based and intentional accounts of privacy that have been addressed so far in this book. In doing so it indirectly denounces those conceptions of privacy that comprehend it in terms of stocks, reserves, property and colloquialisms such as 'protecting my privacy' as this is to use the language of things that have location and extension, thereby falling into a *substance-based fallacy*. While clearly people do not really see privacy as a material object, the use of such language hides the conceptual fact that privacy is better considered in terms of relations, processes, events, emergence, outcomes, consequences and redefinition. As argued throughout, privacy situations arise through interaction, and the word 'action' within 'interaction' is something of a giveaway as to the dynamic nature of privacy. In a people-based sense of privacy, we see movement between social contexts or active management in terms of withholding and disclosing. Correspondingly, in an informational sense, privacy is less about a lump of information residing somewhere, than the circulation of information between interacting data managers, users, hardware, applications and other data sets. Unpacking the dynamic approach to privacy, this chapter explicates in greater detail the notion of events in Whiteheadian parlance. It also highlights the emergent nature of privacy that correspondingly connects well with Rorty, Nissenbaum and Latour discussed earlier, who each in their own way see norms as contingent and affective, and that which emerges in reference to actors, actants and context. This is to emphasise the creative nature of affective privacy events as that which generates novel outcomes that go on to play a role in later situations. In addition to the Whiteheadian exploration of privacy events, this chapter also briefly assesses the reach of Whitehead's ideas about process, and how this contributes to understanding the contemporary media environment; and also suggests Whitehead-inspired methods with which to study privacy and media.

Events

Privacy is less a thing, substance, stock or reserve, but an affective outcome realized when privacy protocol is disrupted. The upshot of this upset is an event and bifurcatory point in which both our understanding of protocol comes into relief and protocol itself may be affected, depending on the intensity of the event. In contrast to inert conceptions, it is to see privacy in more dynamic terms. This understanding of events is to merge both the Whiteheadian and Heideggerian worldview. Heidegger's events are those situations in which being is glimpsed and disclosed. Put otherwise, it is the thisness or truth of a given object or experience understood in reference to unfolding from a much larger context of inter-relationships that comprise the event and the subject therein. The Heideggerian take on privacy events has a retrospective quality as the event provides an opportunity for the analyst to see more clearly the protocol that has been breached, along with the means and actants that contribute to both existing protocol and the event that tests protocol and calls it into question. However, more central to this chapter is Whitehead's (1968 [1938]) understanding of an event as less about disclosure and looking backward, than the future-oriented creation of something new that goes on to play a role in future events. It is creative in that process-based sense of the bringing about of new unities and affective outcomes that participate in and impact on later situations. On creativity and emergent happenings that provide outcomes for subsequent events, Whitehead characterizes creativity as 'that ultimate principle by which the many, which are the universe disjunctively, become the one actual occasion, which is the universe conjunctively' and that it 'lies in the nature of things that the many enter into complex unity' (1948 [1933]): 21).

An event is a consequence of history, arrangements, intersecting vectors, affiliations, bifurcations and change that intermingle to form an entity that possesses an identity, that itself acts as a component of later events. For Whitehead these may be large or small, and we can consider both scales in reference to privacy. For example an event might be a nexus or unit of experience where a confluence of processes fraternize towards an affective moment carried over to the future. A silly but embarrassing example is being on a train and stumbling into the toilet where a person is sitting because they have left the door unlocked due to presumably unclear electronic locking instructions. This is not something one would forget in a hurry (I was the stumbler, not the seated). While a highly notable event for the people involved, clearly experiences were different. Localized experience is accounted for in Whitehead (1964 [1920]) who in discussion of 'percipient events' points out that we do not have a meta-view of what *the* nature of the event is. Instead we are both a causal element somehow, but are also affected by the event in different ways (although I imagine the other party also now double-checks the door is locked). At a macro scale, we might use the example of the UK's Tempora surveillance program, managed by Government Communications Headquarters (GCHQ), that was revealed by the 2013 Snowden leaks to have placed interceptors on fiber optic cables to monitor internet traffic, telephone calls, personal internet history, the content of email messages and social media entries. While the consequences of this privacy event are yet to be fully understood, we can still see that privacy is always connected to arrangements, processes and their coherence into affective outcomes that are involved in later personal, technical, political and protocol-based circumstances.

Whitehead is a recent philosopher whose unique approach is far-reaching and comprehensive, and connects well with the meta-argument of this book, which is to see privacy in terms of protocol, process and affect. A broadly Whiteheadian ontology prefers interactionism, becoming to being, and process to static immutable norms. Another formal and analytical way in which Whitehead (1948 [1933]) is useful is that his ideas about process and affect are cosmological, as to an extent are Latourian and Spinozan/Deleuzian conceptions. This means the objective of inquiry is always process, movement, coherence, bifurcation, disruption, change, multiplicity and the means by which what is novel is generated. Similarly, at the heart of any discussion about assemblages, black boxes, anti-substantialism, or relational, contextual and dynamic accounts of privacy, is Whiteheadian logic. A privacy event is a transformational situation in which a unique configuration of entities and circumstances brings about a novel reaction that alters one or more of its constituents. Depending on the intensity and scope of the event, the publicized outcome may affect and influence subsequent privacy matters and events. A privacy event comes to be from a given 'prehensive' arrangement. This word is unwieldy and difficult, and even Whitehead (1997 [1925]) owns up to the fact that it is awkward. However, as to what prehensions are, they are the ways in which elements within an event might be connected with each other so to form a productive entity, or event (Whitehead (1985 [1929]; McStay, 2013a). The theoretical difficulty emerges not from the principle of co-productive arrangements, but the feeling-out, appetitive and non-consciously willful ways in which actants experience other actants, and perceive, feel and take each other into account. It is to refer to the small yet seismically important details that bring constituents of an event together and how in these Whiteheadian 'occasions' actants touch, interact and respond to each other. This is a microsociology of a Latourian and Tardian sort involving not just interactions between people, but actants from all sorts of domains and scales. Whitehead (to complicate things further) dubs the becoming part of this process as private. This sees the becoming part of concrescence as private as a 'private quality [is] imposed on the public datum' (1985 [1929]: 290). Thus the origins of processes of concrescence and what is to become new are public, but the form by which these are arranged is private. This remains private until the process of genesis is complete and what is novel emerges, this then becoming public. Whitehead's approach is echoed later in Deleuze and Guattari who similarly point out that 'The ("public") matter of fact was the mixture of data actualised by the world in its previous state, while bodies are new actualisations whose "private" states restore matters of fact for new bodies' (2011 [1994]: 154). Similarly, on developing the uniqueness of arrangements and outcomes, Halewood and Michel remark that 'this privacy is only a moment, and this moment is that which constitutes its subjectivity' (2008: 38).

An event is an assemblage of prehensive connections, and respective feeling-out of actants of all scales and persuasions so to form a unity with affective character. However we should not confuse the interest in process with only a systems-based assessment. While the study of privacy assemblages and interacting entities therein involves coolly tracing connections that give rise to the event, and making reasoned inferences about how outcomes may go on to affect future situations, there is more to the analysis. Whitehead's account of events and process is also predicated on affect, care, concern and relevance. It is the combining of this with an interactionist and process-based account that makes him particularly interesting. The basis for prehensions and concrescence (the coming into being of something new) is for Whitehead emotional. This is established on the basis of what Whitehead (1948 [1933]) phrases as an affective tone that generates 'concern' for a given object or entity. This emotionality is that which establishes the relevance of things for us and for each other, and for our purposes serve as those sorting principles that bring about the configuration of a given privacy event. Whitehead (1985 [1929]) phrases this as pre-intellectual, but in more tangible terms we can approach this using the Latourian, Spinozan/Deleuzian and phenomenological material that precede this chapter. By means of Spinozan affect we see that privacy matters are highly charged and far from experientially blank, as with dispassionate systemic approaches. In reference to Latour, we recognize that technology is not blank either but a morass (or imbroglio) of scripts, characters, affordances, tendencies, interests and capabilities. Moreover, in regard to phenomenology and machines, we also established they have intentional properties, with things that exist for them, and processes and objects with which they are concerned. The care and relevance that Whitehead speaks of is not cognitive, but is better recognized by what was phrased earlier in terms of carnality. The background context to a given privacy event (whether this involves people, objects, media technologies, or actants of another sort, e.g. a law) is a seething mass of persuasions, temperaments, modes of behavior, qualities, dispositions and personalities all characterized by being affective and of consequence to other actants that comprise the arrangement from which a privacy event occurs.

Making public of private prehensions

As what is novel comes into public being, it takes on qualities of what Whitehead (1985 [1929]) designates as an 'actual entity' (the outcome of an event and process of concrescence). In Latourian terms we can call this an actant. After the birth of the actual entity, the assemblage/entity/actant/black-box becomes public and a possible datum or element in new arrangement of other actants or entities. This might involve new personal security settings, changes in a brand's behavior, new press regulation, amendments to privacy laws and directives, or possibly even more clandestine security operations. Put otherwise, once an arrangement has come together to provide a new entity that is more than the sum of its parts, what is new is thereby public and able to participate in the formation of new arrangements, events, subjects and actual entities. Thus while having local effects (percipient), if the event is joined by other forces and actants, these provide the potential for the event to scale-up, and for it to take on recognizable and durable black-box status so for the event itself to begin to organize its constituent parts. A notable example of the capacity for black-boxing is the 2011 phone-hacking scandal in the UK. Here journalists and a private investigator illegally (under the Regulation of Investigatory Powers Act 2000) accessed phone messages of murdered schoolgirl Milly Dowler, families of dead British soldiers, and victims of the July 7, 2005 London suicide bombings. This initially included a range of actants: the families of those whose voice-messages were accessed, pushy editors, journalists, the private investigator, the phone service provider, their voicemail services, PIN-based access to phone messages and weak phone security and PIN protection (0000 or 1234 are the most popular). While heinous in itself, this black-box was joined and bolstered by celebrity support, revelations of police bribery and a host of other forces that both shaped and were subsequently shaped by the privacy event. As the outcomes of the event became apparent, chronologically this resulted in the closure of the tabloid newspaper *News of the World*, the cancellation of News Corporation's bid to increase majority share ownership of the British telecommunications company BSkyB, the resignation of the assistant commissioner of the Metropolitan Police, and the Leveson (2012) inquiry that recommended a new independent body with greater control over the UK press (although this was then rejected by the Government). Methodologically, in analysing privacy events we can take a given event and assess the ways in which it concresced or came to be. Indeed, Whiteheadian logic is best applied by considering first a given privacy event that we can agree to be affective in some way; second to look at the ways in which those prehended came to be connected with each other so to bring about the event; and third to monitor the singular or multiple outcomes of that event and how this affects individual or multiple actants (whether these be people, machines, software, laws, organizations and so on). This is a task for the Latourian sociologist who understands that events and society is comprised of the full gamut of actants from all sorts of scales. It is also that person willing to trace backwards, sidewards, forwards, along multiple vectors, *and* be able to conceive of the inter-relationships between these. As mentioned earlier, this requires both the capacity to chart lines of influence so building a systemic picture, but also sensitivity to persuasions, affordances, scripts and dispositions of a wide range of actants. Detail matters, although we run into a problem here, particularly given the cosmological reach of Whitehead, as there are no easy borders or means of limiting our analysis (as with Latour's endless regression of black boxes). There are no easy or straightforward answers here as to when analysis of privacy events should stop. We can however take some recourse to the premise that actual events have a public character and are discernable, recognizable and have objective properties.

Process and media

In addition to theoretically contributing to the understanding that privacy events provide outcomes that go on to play a role in later privacy situations, process also applies well to the media environment itself, particularly in regard to data movement and profusion, and therefore questions about informational privacy. In more specific terms, in regard to informational privacy, this connects to the ways in which bits of information are collected, sorted, aggregated and items or patterns deemed of value to the preferences programmed into a given system are put to use, or go on as metadata to play a role in new assemblages. This ongoing process is productive, and such a view of media rejects ontologies, philosophies or worldviews based on stasis (also see Thrift, 2008). On this point, there are textbook lessons for media studies itself from the Whiteheadian worldview, not least the need to more fully address relations, movement, connection, transversality and the ways in which elements of communication programs combine to form larger affective assemblages in excess of its constituent parts. Information now involves unceasing processes and while earlier control systems by the 1930s involved development of insights on citizens (governmental) or consumers (businesses) by means of feedback loops—this information was put to work in a relatively slow fashion (McStay, 2011). Connective networks, and informational and self-perpetuating assemblages, have most certainly existed for some time but ecologically speaking they were sluggish. By contrast, contemporary systems are intensive unending processes of iteration, aggregation and re-aggregation as new information arrives and insights are distributed (be these advertisements, recommendations or indications of terrorist movements and activities).

While flowery language, such as 'philosophy of becoming' or 'philosophy of media ecology' might appear as hyperbole (and possibly clichéd), it properly reflects the media and systemic environment we live in that increasingly involves Bayesian learning, and empathic machines with intentional capacities, and tireless capacity for attentiveness and connection-making. Whatever our feelings may be about such systems (likely ranging from 'much ado about nothing' to 'we've woken up in a Philip K. Dick nightmare'), the emphasis on connections, interchange, flow, trace, co-emergence, transduction and process is why Whitehead is important. He offers the conceptual orientation to appreciate localized, contextual, emergent and ecological ontologies, even if he did not explore the ways in which such theory played out as is being done here. In regard to informational privacy, particularly in the online environment, these are characterized by feedback relationships, behavior change within a given system, and a sense of continual process without rest. Thus, for a specific example, if we were to look at behavioral advertising, we see that although we may be asleep and away from the computer, the externalized subjectivity of millions of people around the world are helping to refine clusters and segments to which we will later add, contribute and help develop as we turn on our web browsers and start perusing the web. This sense of dynamism, assiduous process, continual movement and learning is an important one as it characterizes well the state of contemporary media as one of unrest, connecting, development and empathizing. Minus the empathy component, this is a very Whiteheadian ontology.

Conclusion

If we agree that an emergent account of privacy is preferable to those that employ fixed norms and rigid borders, and that dynamic and contextual accounts have valid attributes, it is useful to understand the underpinning philosophy and logic behind process and events. Whitehead is a useful port of call, not least as he helps articulate that privacy emerges from societies of subjects and objects; that privacy involves specific arrangements that cause events and bifurcatory outcomes; that privacy events creatively affect actants within the event and beyond it; and that as a result of events, new norms or protocol may be established (for better or worse).

CHAPTER FOURTEEN

Community facts

Clearly this has been a wide-ranging book that has tried-for-size very different ethical and epistemological approaches to philosophy, privacy and media. As highlighted in the introduction, other approaches and philosophers might have been broached and it is quite possible the reader might have chosen a different line-up if tasked with outlining a book that explores privacy and media by means of philosophy. However, each approach selected in its own way contributed to the two meta-principles that lay at the center of my own arguments. While I began writing the book with views about privacy and experience of the topic, these principles also emerged in a somewhat iterative and recursive fashion out of dialogue with philosophical literature that broaches privacy matters (directly and indirectly). To restate, these propositions are: 1) privacy should be conceived in terms of affective events; 2) privacy is an emergent protocol that contributes to the governance of interaction among people and objects. As to specific philosophers and traditions broached, these include Greek dualistic accounts because border-based narratives of privacy continue to permeate discourse on privacy, particularly in Aristotle's (1995 [350 BC]) *Politics*. While to an extent Chapter 2 engaged in a degree of ground clearance so to allow for less rigid (binary) and more emergent approaches to privacy, this basic inside/outside or interior/exterior observation is linguistically difficult to escape. However, the limits of this conception is highlighted in later chapters that problematize ideas of private minds and public language (as what is deemed private is constituted by what is public, and that what we take to be private [such as emotions and feelings] is actually more public than first thought). With the subject to an extent turned inside-out, seeing privacy in terms of border-management becomes figuratively difficult. This leads to the conclusion that greater care of what is public is required. Also finding deep origins in Greek and Platonic accounts is mistrust of the wish to be private that leads us to broadly summarize Greek accounts of privacy in terms of both shame and shiftiness. This is especially noticeable in *The Laws* where Plato (2004 [360 BC]) bluntly requires citizens to be straightforward and not shifty (his translated word choice). Liberalism is also indexical with mainstream conceptions of privacy, particularly as expressed in terms of rights and the relationship with the state. It is also here, most overtly in Kant (1983 [1793]), that we see the roots of contemporary insistence on freedom and autonomy. It is in liberalism that we find the linchpin of consent and that area of civic decision-making that remains deeply topical today in regard to both governmental surveillance, corporate data mining and situations where these activities are combined. Indeed, a broadly liberal approach remains the most agreeable, although caveats are required, particularly in regard to both the constitution of the subject and its excessive emphasis on seclusion and avoidance of others. As privacy is an emergent protocol predicted on connection (rather than being alone), greater acknowledgement of community and living with others is required. Autonomy should be maintained as a premise because it is both a dignifying norm and principle of the functioning of privacy protocol, but might be stripped of its isolationist connotations. Instead (as an ideal) we should view privacy in terms of control, negotiation, management and connection with others. Utilitarianism is included because of its intimate connection to liberalism, particularly in Mill's (1962 [1859]) *On Liberty* and his rebellion in *Bentham* against key areas of Benthamite doctrine (see Mill, 1962 [1838] and also Bentham, 2000 [1781]). However, in Bentham-inspired thinking (that has much older origins in Cumberland, 2005 [1672]) we see a problematic dimension for privacy in terms of enlightened interest in both reason and transparency. While generally (and rightly) regarded in positive terms, a question remains over how far transparency should go and when as a guiding social value it becomes perverse. To answer this, I proposed a division between transparency and radical transparency, or that will to open all areas of life for analysis. Is radical transparency a desirable state of affairs? Does the call for greater transparency hide motives other than enlightened interest in the acquisition of knowledge for improvement of humankind? Or, might the will to radical transparency be read in dystopian (yet all too real) terms? Pragmatism, while liberal in orientation, vigorously questions one-size-fits-all deontic norms and Habermasian searches for universal validity, and best explains the recent interest of the privacy community in contextual norms (Nissenbaum, 2004, 2010) and the ways in which social circumstances dictate privacy controls (Altman, 1975; Altman and Chemers, 1980). This connects with my second principle on emergent and dynamic protocol. Imbued within this is another basic but important point that privacy far exceeds our current interest in media. I argue that as an emergent protocol or co-authored script that guides and organizes principles of interaction between actors, privacy is very much connected to recent concern about mediated communication, but it is also older, more basic, and one might even venture primal and biosemiotic. On the point about protocol and principles, pragmatism is a creative philosophy in that it does not seek to preserve norms but rather its ethnocentricity and emphasis on contingency allows privacy protocol to develop as its members see fit. While the emphasis on contingency and localism is important, what is not up for discussion is the existence of privacy as an affective set of protocols that guide social arrangements in respective social groups (be these globally dispersed or within an individual household). Moreover, while one can see how local preferences versus one-privacy-size-fits-all has attraction, I have reservations about introducing this into media policy as there remain unanswered questions in contextual accounts regarding consent, and the means by which we are meant to be aware of the vast array of privacy protocols we find ourselves involved with and immersed in. Until these involve explicit consent and not tacit or assumed consent, and means for properly negotiating the principles of protocol with actants, norms must remain as to move towards jettisoning these leaves us open to redefinition of contextual principles by resource rich organizations.

Heidegger provided two key contributions: first his idea that technology is not especially technical (see 1993 [1954]); second, and less well-known, is his emphasis on events as that which discloses the truth of being (see 2013 [1941–2]). The first point is an important one as it helps place contemporary concerns about the nature of data mining technologies into a much larger continuum of an unfolding rationality that seeks to rationalize, and make present-at-hand and usable. This observation about technical being and the sense that technicity transcends our immediate period is useful as it also recognizes the semi-autonomous nature of our technical environment. This is less about determinism, but recognition that we (and technologies) are born into already-technical domains. While we *are* able to effect change to our domains, our technical environment clearly has separateness because of the cumulative effects of rationalist metaphysics and technological history. We perhaps most keenly feel this now in the will-to-make transparent evidenced by recent accelerated intensification of security and marketing endeavors. Beyond privacy, for Heidegger (2011 [1962]) something important is lost in this will to make all areas of humanity and life into a reserve. In being technical, being itself is forgotten, along with that metaphysical and phenomenological interest in experience that surpasses properties, attributes and affordances. This is vague, but being is that which transcends the object itself and provides us 'giveness,' 'that it is' or the 'whatness and thisness' of experience. While there is quasi-romanticism to this, within Heidegger's obscure writing are useful observations about how things come to be. Part of this is standard phenomenology on structures of subjective experience, but Heidegger also introduces greater separateness of objects and their backgrounds, and the totality of relationships and contexts that give rise to them. These observations allow for greater interaction between ourselves, being-in-the-world and co-created moods that generate what are disclosed to us. Applying Heidegger to networked privacy matters, it is to study how the beingness of our heterogeneous informational environments come to be the way they are. This carries us into his second contribution, that an event in Heideggerian discourse is the disclosure of the suchness of things. Heidegger's use of the word 'event' is different to my own (mine is more Whiteheadian in orientation) and while highly important for Heidegger, this latter point may appear for the reader a confusing and possibly banal observation about what it is to know things around us. The value of the discussion, however, becomes clearer and more interesting when we start to ask questions about whether machines and behavioral watching systems are able to engage with the various 'nesses,' and phenomenal ways of knowing and discerning. Heideggerian philosophy thus leads us into questions about knowing, verisimilitude, representing and the capacity to engage and interact with human phenomenal being. His worldview is worth resurrecting because while we are safe for now in our assumptions that machines do not understand being-in-the world, what is less understood are the consequences of representional, Bayesian and empathic processes, and their increasing capacity to preempt and map being-in-the-world—evidenced by capacity for prediction and going unnoticed.

The interest in technology continues in Latour (1987, 1992, 1993, 1996, 2004), although Latour's approach is somewhat different and more positive than Heidegger's. Latour raises the status of objects to actants and provides them greater social recognition because of both what they allow people to do (human history does not do justice to the objects that facilitated it), but also because of the affordances, properties and capacity to affect that technology represents. Latour's (2005) microanalysis of actants is useful for privacy matters because it reveals a porous interchange between categories, processes and levels of arrangements. To properly study privacy requires some very unglamorous yet revealing analysis of accounting and management practices, economics and business, governmental decision-making, engineering, law and regulation, informational management and statistics, marketing and advertising, usability testers and many other areas that contribute to the roll-out of hybrid actants such as smartphones, social networks and surveillance systems. The reason for doing this is that there is much interaction and influence between these scales and zones of organization that is highly revealing in trying to understand a given privacy matter (for example cookie use and behavioral advertising). Discussion of Latour allows us to develop the idea that while privacy is not a thing, substance or object—it does exist. It also links very clearly with the first proposition that privacy is well thought of as an affective event. Privacy emerges from a political ecology of actants and exists as an outcome. It is granted the right to exist because it is capable of affect, whether this be in terms of policy change, the altering of software and hardware, or simply refusing to speak to someone because they disclosed a secret we entrusted that someone with. Privacy thus becomes an actant in its own right and modifies and affects that which is around it.

I continued to employ continental philosophers by means of returning to phenomenology, particularly Brentano (1995 [1874]) and Husserl (1970 [1900], 2002 [1952]), and their interest in the in-existent dimension of things. In the chapters on Heidegger I laid the foundations for a discussion of empathy and the ways that increasingly technology seems not only to be watching us, but also to know and respond to our preferences. While we rightly say that human knowing and machinic knowing are quite different, it is not entirely clear why. While we are certainly more complex we engage in many of the same intentional processes of representing, reading and responding, and importantly we also make mistakes and misread situations. Further, on intentionality, it is not clear why either intentionality or empathy should be denied to watching and responding machines. After all, if intention is that which structures experience (Husserl, 1970 [1936]), can we begin to see such a process affectively taking place at all levels of reality? This idea of ascribing intentionality to objects and the cosmological/Whiteheadian shift in gear it invokes belongs to Harman (2005), but in regard to technology and privacy it means we look at machines differently as we recognize them as that which are attentive *to* something. As machines become intentional and empathic, their status is raised somewhat. Empathic character is a key argument of this book and if the reader can agree that empathy and machinic intentionality better characterizes behavioral technology than 'intelligence,' then this book will have fulfilled one of its objectives.

After focusing on technology and machines involved in privacy matters, it was important to consider in further depth the objects of empathic media, i.e. us. My scrutiny of the subject made some familiar and unfamiliar observations. In general, what was found is that in many ways we are much more public than a liberal account of the self leads us to believe. The stable sense of a willing *I* becomes problematic when we ask questions about the language we use and the impossibility of expressing a truly private experience (Wittgenstein, 2009 [1953]). With these observations on the publicness of *I*, its (our!) assured sovereignty is diminished somewhat. Moreover, while we do not have to accept the full consequences of behaviorism, there are lessons to be learned on public/private distinctions in regard to subjectivity, not least Ryle's (2000 [1949]) message about dispositions, the ghost in the machine and his insistence that mind is not something behind the behavior of the body, but rather is part of the behavior of the body. This also witnesses people as living more publically than an introspective mentalist approach suggests, and opens up the subject hitherto conceived as private to public scrutiny. We do not necessarily need neuroscientific equipment to do this either as much of this involves movements, utterances, noises, temperature, chemicals released, interaction with environment, choices and so on. In the Skinnerian (1976) sense, thought is behavior. This is to turn border-based conceptions inside-out so when we talk of feelings and propensities (being kind, angry, sad or shrewd), these are not interior qualities but public behavioral traits. This attack on supposed private interior theaters of the self is less about reducing the value of the subject, but a privileging and accentuation of our public selves. This is a key point because by over-accentuating the phantom subject, we take critical attention away from the real public exterior self that provides a far better record of what and who we are.

The consequence of this making-private-public is that empathic machines do not have to access a special interior realm to get at the real us, as we freely parade interiority publically. Empathic machines do not have to feel-into, they simply have to respond appropriately (the same goes for empathic people too). The subject then is very much public and the private *I* is vastly reduced as we understand that the self comes to be by means of symbiotic reciprocation. This is not to deny entirely the decision-making self (we can withhold, choose and engage in willful acts) but it is to observe that the liberal self is not an isolated being and is deeply connected, communal and ecological at a number of levels (most relevant here being language, culture and belonging to social contexts). A key outcome of this argument involves a reduced subject and the observation that much of what we thought to be interior (private) is actually exterior (public). This is of consequence because to not notice this is to run the risk of over-privileging a mythical interior subject and concurrently de-value our real exterior selves. It is also to fall prey to that liberal construction of the monadic subject when in truth, subjectivity comes to be by means of public behavior. Thus in addition to the under-prizing of the public self, we have also under-valued and misrecognized the principle of relations and connection. With the self more public than private, I argue that greater care is required over what is public. This takes on renewed importance in an age of empathic machines and media that is based on the fact that what is public is valuable and knowable, and to over-invest in the monadic subject is to de-value what is public and handover our real exteriorized selves.

An account of philosophy and privacy that involves media cannot ignore Marxist-inspired critical theory. Privacy studies in general tend to be of a critical orientation (this is why Richard Posner, discussed in Chapter 4, stands out so distinctly). However, Marxism sits uncomfortably with privacy because of the problem of property and ownership. It does not look favourably on discussion of rights or autonomy either because of the deep emphasis of liberalism on seclusion and separation (Marx, 2012 [1844]). This weighting towards individualism is, for Marxism, antithetical to the emancipatory project that emphasizes community over isolated monads. More recent discussion, particularly within the Autonomist Marxist community (I have in mind Franco Berardi, Michael Hardt, Maurizio Lazzarato, Antonio Negri and Paolo Virno), has focused on the immaterial dimensions of contemporary labour (Berardi, 2009). Much in critical new media studies has been made of this in regard to Smythe's (1981) audience-as-commodity discussion (also see McStay, 2011), but a key question remains—is fidgeting, swiping and tweeting with smartphones, tablet and other computers *really* labor? This is a very basic question I ask my Year 1 undergraduate students, but it is one still requiring a good answer. My response is 'sort of, but only if we accept that labor is defined through value generation for another.' One might question what Marx and Engels (2011 [1844]) would have thought of this weak answer. Rather than stretch labor to the point of being unrecognizable, I suggest a slight change of critical emphasis. More productive and illuminating is the principle of alienation. Found respectively but differently in Hegel and subsequently Marx, they both involve a schism between the self and the world. An interesting question is raised if we agree with my argument from the previous chapter that the self is more public than first thought, the interior space for a subject is small (if it exists at all), and what we display publically by means of behavior and connections with others is valuable because, essentially, that is what we are, i.e. that which comes to be by means of connection. My argument and recommendation in regard to critical theory is that if we are to take lessons from Hegel and Marx we might pay less attention to online media use as immaterial labor. Instead we see that consciousness of alienating processes entails attentiveness to the value of public traces because with the bubble of the subject popped, greater value should be placed on what is public. This progresses us to a bizarre observation for a book about privacy: to an extent privacy is a fallacy and what is required is greater control of what is public. While I do not agree that a Marxist society is desirable (particularly because of privacy consequences, the means of enforcing community norms, and the unlikeliness of achieving a better society than the one we already have), the theoretical observation of processes characterized by alienation is a useful lesson to take away from Marxist theorizing.

Next in the line-up of my choice of philosophers was Spinoza. He is employed primarily to counter the sense that privacy is an abstract and solely theoretical notion. Much has been made in this book of systemic, contextual and dynamic privacy protocols and although affect is addressed in discussion of Latour, it is Spinoza (1996 [1677]) who tells this story best and helps us establish affect as that which involves mind and body, and that these are ultimately of the same substance. Deleuze (1988 [1970], 2011 [1969], 2011 [1981]) also assists with helpful re-readings of Spinozan principles of affective bodies and that which is able to be designated a body because of the capacity to affect and be affected. By means of Spinoza we can de-intellectualize privacy and more easily consider consequences of privacy events in terms of immanence and corporeality. Ultimately, privacy is not a theoretical proposition, but it is lived, felt and affects at a range of scales. However, we need not restrict ourselves to only the ways in which we may be affected, but we can also see the affective properties of privacy protocol. Sartre (2003 [1943]) depicts this with clarity as he regales the story of the voyeur who became an object for either a real and imagined Other (it doesn't matter which, it is the point about privacy protocol and the surveiller affect that counts). Finally I invoked Whitehead (1985 [1929], 1948 [1933], 1968 [1938]) because of his far-reaching account of context, affect, events and the ways in which outcomes influence later events. Whitehead is important in regard to both the emergent and affective propositions developed in this book. Very much under-recognized, his worldview directly informs any commentary on media, philosophy or indeed privacy that privileges process, connections, transduction, interchange, flow, trace and co-emergence over substance and rigidity. His account of a creative cosmology argues that at the most fundamental level newness is possible, and that what emerges out of bifurcatory events go on as actants to play roles in later events.

Community fact

Does privacy have a place where it resides? No. Does it come to be by means of interaction among a wide range of actors? Yes. Is it relativist? To an extent. While privacy does not have substance, it is quite real, and when we say real, we are able to say that it can bring about corporeal, behavioral, psychological and organizational differences. Given the reach of privacy as a regulatory tool across a wide spectrum of entities, and that it is a governing and therefore political norm, this grants it parental status in the sense that it is an old, common and affective premise that precedes many recent rights. It has a primal and fundamental character, and while in the hierarchy of rights it may be trumped, as a *community fact* we do not have to argue for its existence or articulate why it matters. In Chapter 4 I mentioned Rosenberg's (2000) attack on privacy that sees it as relativist, lacking commonality and akin to individual taste (a moral truffle). Privacy as an emergent protocol *is* ethnocentric and relative, but as a principle that involves establishing norms of our interaction with others (be these people or machines) it seems to be universal.

With privacy being affective protocol, the ethical onus for anyone seeking to modify or alter it is to make the case for their actions and obtain full consent by community members. To do otherwise is an act of force. Privacy is much farther-reaching in scope than we may have hitherto thought. While academic and policy scrutiny has focused on informational privacy and the impact of novel technologies and their uses, privacy reaches into the details and micro-interactions of everyday life. While concern about consent, cookie use, legislation and the involvement of corporate and governmental surveillance in our mediated communication flows remain critical areas of current consideration and scrutiny, these topics are renewed and refreshed by recognition of the breadth of privacy matters. This breadth is comprised of the fact that privacy, as protocols and ethnocentric emergent norms, is found in the most seemingly basic of arrangements. It is that which modulates and informs the principles of connection with others and on this point we can agree with the Marxist premise that privacy is not about isolation and the monadic subject, but connection and communal norms. This bottom-up observation is an important one because it highlights the wrong-headedness of privacy as being somehow lost because means of telecommunication have expanded. Rather, the more basic observation that privacy is an emergent protocol that mediates relations between people and technical systems promotes the premise that privacy is a fact of communities.

Despite any possible attempt to escape liberalism by recourse to other philosophical worldviews to explain privacy, the gravity of liberalism cannot be eluded. They are inimically tied together and given that pragmatism (whose self-stated *raison d'être* is to look forward rather than backwards) cannot escape the gravity of liberal conceptions of privacy, we will not either. However, we can modify the liberal outlook so that privacy can be understood in terms of protocol and can be collectively (rather than individually) cared for. We can build, expand, develop, criticize, take to task and contextualize—particularly by means of pragmatism itself. Most notable is the need to admit that rights are relational and dynamic so to be able to admit of difference, deviation, dissent and diversity. This sees rights not as inherent properties of individuals, but as a matter of relations between consenting agents. This leaves privacy up for renewal although citizenry need to carefully consider consequences about the nature of consent, questions about understanding the terms of situational protocol, and the resources companies and governments have at their disposal to achieve aims of increasing levels of access to information.

Adopting a pragmatic outlook

In considering privacy as that which is an affective outcome of ethnocentric protocol breaches, we find ourselves erring towards a more pragmatic outlook. Privacy is not dead, a defunct norm, kaput and we should not 'get over it,' but it is being contested by powerful actants. However, by better understanding how very basic and fundamental privacy is, and how far protocol affectively extends into our lives (far exceeding computer networks), we are able to put glib assertions into place. Recognition of the reach of privacy into our lives and the means by which protocol affects relations between each other places us in a stronger position both to weigh-up our own personal position on information, privacy, marketing and security. It also allows us to see attempts to reduce privacy as highly ideological, particularly in regard to radical transparency. In adopting a pragmatic view of privacy we remain open to the fact that it is open for redefinition, but such contingency invites critical appraisal of the motives of those who seek to redefine privacy protocols.

Nonetheless, a pragmatic approach attempts to avoid the deductive principles and assumed truths of much critical theory. This is less about being apolitical, but rather a wish to proceed on the understanding that specifics of local and immediate situations count for more than ruling norms drafted in from past epochs. While recourse to historical influences (both material and metaphysical) are to be made, these traces and associations need to be made explicit, and their relevance argued and explained rather than assumed. So, for example, on complaining about the potential harm of gathering information, consequences need to be spelled out with as much clarity as possible. Arguing that a process is morally wrong (from the perspective of Marxism, deontology or any other normative worldview) is not enough. Consequences need to be brought to the fore and accounted for, and implications of processes need to be articulated. Theory and cultural norms matter, but only as historical learning combined with an assessment of the contemporary specifics of law, corporate and governmental decision-making processes, geo-political and security pressures, technical design, machinic affordances, hybrid actants and the current preferences of people, who each contribute to privacy protocols (whether this be of a small local culture [a family, workplace or restricted online group] or continent-wide protocol, e.g. European policy on cookie norms).

Affective privacy

While privacy has primitive and pre-theoretical qualities, mediated privacy involves additional layers of complexity difficult for even the most technologically proficient people to manage—often because of the limited choices a system might offer. However, my interest has been to build an approach to privacy that is broader than an information-focused account as the majority of situations involving the possibility of privacy events tend not to involve computer networks, but conversations with colleagues and partners, the distance we sit and stand from each other, doors and architectural norms, and other seemingly mundane day-to-day occurrences. Privacy protocol (as an outcome of past privacy events) regulates and affects much of everyday life. This affective field that in general we abide by without realizing is paramount because affective protocol both emerges from the minutiae of everyday practices and regulates them. This very much connects with Heidegger as to understand the emergence of privacy protocol in any given context is to ask questions about being, i.e. to investigate hermeneutic contexts and the social horizon that provides reason and intelligibility for our behavior. This is not simply social and cultural in that breezy abstract sense in which we use these two words, but it is also visceral in the sense that protocol becomes embodied and part of the corporeal 'us.' This can be witnessed both in terms of behavior and when protocol is breached, potentially giving rise to visceral reactions. To study privacy protocol is to inquire about its ethnocentric being and means of emergence.

The code and being of such protocol (in the Heideggerian sense of understanding context) comes into relief when protocol is broken and a privacy event occurs. This might involve finding out information has been shared with the wrong person, realizing we have been tagged in particularly bad photos online or, more seriously, on suffering a domestic or bodily violation of some sort. The disclosure dimension involves insight into the political ecology that gives rise to privacy protocol, and the affective aspect is caused in part by deviation from this protocol. In considering systems of associations that both contribute to privacy protocol (norms) and events (breaches), Latour's (1999) idea about black boxes has bearing. This is because the presentation of a world without substance grants entry to reality of all sorts of actors that are abominations to materialists. If we remember there is no ultimate 'stuff' behind a black box, but only more black boxes, this leads to what Harman (2009: 106) phrases as metaphysics of 'infinite regress.' We should be clear that this is no postmodern shenanigan (Latour is acerbic about Kantian-inspired postmodern goings-on[1]), but a serious attempt to open the terrain of discussion of what actually exists and to eradicate the border between nature and artifice. It is serious because it is methodological and involves studying actors and processes operating at substantively different scales, yet are able to contribute to the emergence of privacy protocol and likewise be affected by privacy events. Despite the intricacies employed to explore privacy, it is exceedingly basic and very real. While the content of this book has in parts been highly abstract and not always straightforward, it has had as its referent a very visceral sense of privacy as that which guides, steer and orients behavior. While privacy is a norm and principle both in moral and systemic terms, it is felt and lived. Privacy is very much part of the human equation and suggestions that it might be waning are to be treated skeptically. However, as that which mediates between our public selves and our technologies, we should be vigilant to the means and motives of those who seek to alter and make use of these connections.

APPENDIX

An A to Z of privacy: New theories and terminology

- *A-historical data mining*: the capacity for equal recall to either recent or older behavioral traces.
- *Affective breach*: that visceral sense of discord that occurs when undergoing a privacy event and when privacy protocol has been contravened.
- *Body-doubles*: an inversion of 'data-double' that is less interested in our mediated self than the impact of mediated behavioral traces on our corporeal self.
- *Border-based privacy*: a foundational binary principle of privacy (i.e., in/out). While easily criticized it continues to permeate discourse on privacy.
- *Co-evolving authorship*: the ways in which by means of interaction with machines we collaboratively publish our own heterogeneous media content.
- *Community facts*: an observation about privacy that recognizes both its ethnocentricity *and* its universality, and involves the identification of the nature of guiding principles and protocol that contributes to the norms of interaction with others (be these people or machines).
- *Dispositional competence*: the test to ascertain that the principle of *machinic verisimilitude* (outlined below) is correct and that machines can pass-off human understanding and predict our forthcoming behavior.
- *Empathic machines/media*: the capacity for near and remote machines to pick up on the emotional state of people, their intentions, their expressions and actions, along with behavioral cues, and act on them.
- *Events*: circumstances involving an affective breach that, depending on severity, may give rise to outcomes that play a role in later situations involving privacy. These might involve

interpersonal relations, the design of technological systems, law-making and regulation, or surveillance practices, among other possibilities. Privacy events also bring *privacy protocol* (accounted for below) into greater view by means of better comprehending the being of protocol, i.e. the regulatory norms that organize a nexus of actants.

- *Everyday mutuality*: that view of privacy that adopts the liberal emphasis on privacy and dignity, but places less emphasis on seclusion and greater emphasis on modulation of openness to one's community.
- *Local realism*: an expression used to make the point that although contextual privacy does not involve universal norms, the affect of such privacy protocol on local actants is very real, affective, discernable and able to be studied.
- *Machinic intentionality*: refers to objectifying processes of machines. It is the means by which watching machines discern an object of sorts, are attentive, have a focus and tend towards a thing to which their attention is drawn. Intentionality is usually ascribed to people and refers to that in-existent dimension of a thing (e.g. mugness, iPodness or pencil sharperness), but in this machinic context involves broader non-human objectifying processes. The existence and capacity of machinic intentionality is a critical component of *empathic machines/media* as outlined earlier.
- *Machinic verisimilitude*: involves a semblance of machinic knowing of nuances and subtleties of qualia and human experience. It is based on the competences and dispositions of machines that interact with granular human experience. It thus involves the co-creation of experiences correct for a given situation (also see *co-evolving authorship*). Veracity is judged by the predictive capacity of machines, for example in terms of what we might click or purchase.
- *Mood of information*: this is the tone of interaction co-created by people and technical media systems. Conceived in relation to advertising experiences online, the expression derives from Mark Poster's early work on post-structuralism and data mining, and Heidegger's on moods as that which are the vehicle that discloses things.
- *Moods*: involve quantitative technologies engaging with qualitative being and behavior. They are a co-authored outcome and heterogeneously created (also see *co-evolving authorship* earlier). To study moods in networked media is to examine the Heideggerian premise of why things are disclosed to us the way they are, and how the tone and beingness of our environment comes to be the way it is.
- *Oppressive autonomy*: refers to that dimension of autonomy that assumes we are able to make unfettered decisions in every situation. The 'oppressive' prefix refers to the fact that this is exceedingly difficult, if not impossible, and while autonomy is positive in intention, it may stand against us.
- *Phenomenal materialism*: refers to affective experience. While privacy involves the specifics of arrangements on how information, objects, territory and bodies should be managed by ourselves and others, a systemic analysis of these arrangements runs the risk of misunderstanding privacy because of a tendency to abstraction. Phenomenal materialism is employed to emphasize the experiential (phenomenal) and physical feeling of privacy (corporeal).
- *Political ecology of privacy*: involves tracing the population and citizenry of situations involving privacy protocol. The words 'political' and 'citizenry' are employed to aid in recognizing that machines and their capacity to influence privacy protocol grants them

a higher status in social arrangements, and that society is not restricted to people. This does not involve the ascription of liberal rights, but rather greater recognition of the capacity to affect situations and outcomes.
- *Privacy protocol*: refers to what otherwise might be phrased as privacy norms. The word protocol is employed so to avoid the moral dimension and discussion of privacy as a normative action, and to draw attention to protocol as that affective 'force' that steers, orients and guides behavior. Like gravitation, it is that which may guide the behavior of actants but of itself be unseen.
- *Privacy scripts*: these contribute to the terms and parameters by which actants of a system interact and behave. A system may involve human-only, human–machine and machine–machine arrangements.
- *Public traces*: are trails of our behavior that are prone to being instrumentalized, used for the gain of others and possibly even against us. This may involve human–human as well as human–machine situations.
- *Radical transparency*: refers to that Enlightenment tendency to unconceal, and make available and present so to be used or employed in some way. While transparency is positive in regard to uncovering machinations of state secrecy and power, questions are raised about whether radical transparency and total openness is a desirable vision for society. It is that perversion which opens up both public life and those of citizens for unwanted inspection.
- *Seclusion trap*: while privacy involves respite from others (be these people or networked systems) it is also that process by which we might also make ourselves more open to others. It is a process of managing connections and is less about solitude and being alone.
- *Substance fallacy (of privacy)*: a fallacy of category type, this is apparent when people refer to privacy as a stock or reserve that can be given away or exchanged. The substance fallacy is not only a problem of language and category, but as privacy is ascribed thingliness it also tends to be connected with property-based conceptions and placed within the domain of exchange. This is to invoke another fallacious treatment of privacy, not least because the privacy-as-commodity argument falls apart in the face of state surveillance matters where free and rational exchange is not an option.
- *Surveiller affect*: the experience of a voyeur existing for a real or imagined Other. It is the anxiety of the watcher who most fears that their watching may be made public. It also involves the collapse of the watcher's world if caught by a real Other.
- *Technically negotiated privacy*: this refers explicitly to interactions with technology by either people or other technical artifacts. It recognizes that privacy is not just about human privacy norms (as with liberalism), but also the need to negotiate with objects and systems. As such, privacy not only has an ethnocentric nature, but also a technocentric nature (although each informs the other).
- *Third-person being*: the state of being 'always-on' and living with the gaze of the Other. Surveillers may be actual and close; mediated and possibly involving technical non-human actors; or even imagined.
- *Zombie media*: equates to non-human empathic actants that can answer questions, provide appropriate responses, predict preferences and outcomes, offer verisimilitude of knowing, yet not possess what in common sense terms we refer to as consciousness.

Notes

Chapter One: Introduction

1. We might also recognize the domestic abuse of men in the private domain. Data from the Home Office and British Crime Survey show that men comprise 40% of domestic violence victims each year between 2004–05 and 2008–09 (see Home Office, 2011).
2. I know there are key approaches missed but if you would like to contact me with suggestions and sources, please feel free to do so, or send me your own work that uses philosophy as a means of broaching privacy (I would very much like to read it).

Chapter Two: Aristotle, borders and the coming of the social

1. Plato is not explicit on what else the newlyweds might be doing, but the women responsible for informing 'should report to her colleagues any wife or husband of childbearing age she has seen who is concerned with anything but the duties imposed on him or her at the time of the sacrifices and rites of their marriage' (2004 [360 BC]: 222 §6.784b).
2. In particular his disclosure about the National Security Agency's (NSA) PRISM operation that collects information from emails, online chat, videos, photos, stored data, file transfers, notifications on target activity, social networking details and other special requests that might be made to internet companies.

Chapter Three: Liberalism, consent and the problem of seclusion

1. On foundational human goods Allen (2011: 13) appropriates Rawls' notion of 'primary goods' to make the point that choice needs to be constrained so to facilitate a 'lifetime of self-respect, trusting relationships, positions of responsibility, and other forms of flourishing.'

Chapter Four: Utilitarianism, radical transparency and moral truffles

1. Berlin (1969) discusses at some length J.S. Mill's childhood and background remarking that both James Mill (J.S.' father) and Jeremy Bentham were both directly severe utilitarian influences on Mill. Indeed, from the earliest of ages he was schooled in ultra-rationalism although Mill in his later teenage years would break away from these.

Chapter Five: Pragmatism: Jettisoning normativity

1. Although it is notable that in a social media context few of us are adept at audience management, with most of us having a poor grasp of the scale and composition of our audience in a social media context (Litt, 2012).

Chapter Six: Heidegger (Part 1): Concerning a-historical being and events

1. See for example Heidegger (2012 [1936–8]: 12).

Chapter Seven: Heidegger (Part 2): On moods and empathic media

1. Created by John Searle, The Chinese Room argument disagrees with the possibility of true artificial intelligence. It is premised on a hypothetical thought experiment in which someone who knows only English sits alone in a room following English instructions for manipulating strings of Chinese characters. Searle's point is that the English speaker is able to operate the symbols in such a way that the Chinese speaker is convinced the 'someone' can actually understand Chinese.

Chapter Eight: Latour: Raising the profile of immaterial actants

1. The irony is that Schmidt and Cohen are Eric Schmidt (President of Google) and Jared Cohen (Director of Ideas at Google) who both regularly find themselves embroiled in accusations of designing increasingly invasive technologies and processes.
2. To slightly amend the title of Bennett's (2010) book *Vibrant Matter: A Political Ecology of Things*.

Chapter Fourteen: Community facts

1. On the link between Kant, postmodernism and philosophy (and forever being stuck in the phenomenal realm), Harman summarizes Latour by remarking, that postmodern philosophy 'holes up in the human castle while disputing whether it has the right to leave, with a few hotheads claiming that there is nothing beyond the castle in the first place' (2009: 148).

References

Allen, A.L. (1988) *Uneasy Access: Privacy for Women in Free Society*. New Jersey, NJ: Rowman and Littlefield.
Allen, A.L. (2003) *Why Privacy Isn't Everything: Feminist Reflections on Personal Accountability*. Lanham: Rowman & Litterfield.
Allen, A.L. (2011) *Unpopular Privacy: What Must We Hide?* Oxford: Oxford University Press.
Altman, I. (1975) *The Environment and Social Behavior: Privacy, Personal Space, Territory, Crowding*. Monterey, CA: Brooks/Cole.
Altman, I. and Chemers, M.M. (1980) *Culture and Environment*. Belmont, CA: Wadsworth.
Andrejevic, M. (2007) *iSpy: Surveillance and Power in the Interactive Era*. Kansas, KS: University of Kansas Press.
Arendt, H. (1998 [1958]) *The Human Condition*. Chicago: Chicago University Press.
Aristotle (1995 [350 BC]) *Politics*. Oxford: Oxford University Press.
Aristotle (1998 [350 BC]) *The Metaphysics*. London: Penguin.
Aristotle (2008 [350 BC]) *Physics*. London: Penguin.
Aristotle (2009 [350 BC]) *The Nicomachean Ethics*. Oxford: Oxford University Press.
Arvidsson, A. (2004) 'On the "Pre-History of the Panoptic Sort": Mobility in Market Research' *Surveillance and Society*, 1(4): 456–474.
Bakir, V. (2010) *Sousveillance and Iraq: Impact of Emergent Web-based Participatory Media on Strategic Political Communication*. New York: Continuum.
Bateson, M.C. (1991 [1972]) *Our Own Metaphor: A Personal Account of a Conference on the Effects of Conscious Purpose on Human Adaptation*. Washington, DC: Smithsonian Institution.
Bateson, G. (2000 [1972]) *Steps to an Ecology of Mind*. Chicago: University of Chicago.

Beccaria, C. (1767) *An Essay On Crimes and Punishments*. New London: J. Almon.

Benhabib, S. (1993) 'Feminist Theory and Hannah Arendt's Concept of Public Space,' *History of the Human Sciences*, 6(2): 97–114.

Benn, S.I. and Gaus, G.F. (1983) *Public And Private In Social Life*. London: Croom Helm.

Bennett, C.J. and Raab, C.D. (2006) *The Governance of Privacy: Policy Instruments in Global Perspective*. Cambridge, MA: MIT.

Bennett, J. (2010) *Vibrant Matter: A Political Ecology of Things*. Durham: Duke University Press.

Bennington, G. (2011) 'Kant's Open Secret,' *Theory, Culture & Society*, 28(7/8): 26–40.

Bentham, J. (1834) *Deontology: Or, the Science of Morality*, ed. J. Bowring. London: Longman, Rees, Orme, Brown, Green & Longman.

Bentham, J. (1843 [1792]) Anarchical Fallacies; Being an Examination of the Declaration of Rights Issued During the French Revolution. Republished in J. Bowring J, ed. *The Works of Jeremy Bentham, Vol II*. Edinburgh: William Tait, http://oll.libertyfund.org/title/1921/114226 on 2013-03-26, date accessed 26/03/13.

Bentham, J. (2000 [1781]) *An Introduction to the Principles of Morals and Legislation*. Ontario: Batoche Books, Kitchener.

Berardi, F. (2009) *The Soul at Work: From Alienation to Autonomy*. Los Angeles, CA: Semiotext(e).

Berkeley, G. (1988 [1710]) *Principles of Human Knowledge and Three Dialogues Between Hylas and Philonous*. London: Penguin.

Berlin, I. (2006 [1958]) 'Two Concepts of Liberty,' in D. Miller, eds. *The Liberty Reader*. Edinburgh: Edinburgh University Press. Pp. 33–57.

Berlin, I. (1969) 'Introduction' in *Four Essays on Liberty*. Oxford: Oxford University Press.

Bermejo, F. (2009) 'Audience Manufacture in Historical Perspective: From Broadcasting to Google,' *New Media and Society*, 11(1&2): 133–154.

Bijker, W.E. and Law, J. (1992) 'General Introduction,' in W.E. Bijker and J. Law, eds. *Shaping Technology Building Society*. Cambridge, MA: MIT Press. Pp. 1–14.

Birchall, C. (2012) 'Introduction to "Secrecy and Transparency": The Politics of Opacity and Openness,' *Theory, Culture & Society*, 28(7- 8): 7–25.

Bobbio, N. (1989) *Democracy and Dictatorship*. Cambridge: Polity Press.

Brentano, F. (1995 [1874]) *Psychology From an Empirical Standpoint*. London: Routledge.

Briggs, A. and Burke, P. (2009) *Social History of the Media: From Gutenberg to the Internet*. Cambridge: Polity.

Callon, M. (2012 [1987]) 'Society in the Making: The Study of Technology as a Tool for Sociological Analysis,' in W.E. Bijker, T.P. Hughes and T. Pinch, eds. *The Social Construction of Technological Systems: New Directions in the Sociology and History of Technology*. Cambridge, MA: MIT. Pp. 77–104.

Cameron, D. (2011) 'David Cameron: We are Creating a New Era of Transparency,' *The Telegraph*, http://www.telegraph.co.uk/news/politics/david-cameron/8621560/David-Cameron-We-are-creating-a-new-era-of-transparency.html, date accessed 29/10/13.

Churchland, P.M. (1990 [1984]) *Matter and Consciousness*. Cambridge, MA: MIT Press.

Clarke, S. (2013 [1728]) *A Discourse Concerning the Being and Attributes of God, the Obligations of Natural Religion, and the Truth and Certainty of the Christian Revelation*, http://archive.org/details/discourseconcern00clar, date accessed 20/08/13.

Cohen, J.E. (2012) 'Introduction: Imagining the Networked Information Society,' *Configuring the Networked Self: Law, Code, and the Play of Everyday Practice*, http://www.juliecohen.com/page5.php, date accessed 10/12/12.

Coplan, A. and Goldie, P. (2011) 'Introduction' in A. Coplan and P. Goldie, eds. *Empathy*. Oxford: Oxford University Press. Pp. ix–xlvii.

Critchley, S. (1996) 'Prolegomena to any Post-Deconstructive Subjectivity' in S. Critchley and P. Dews, eds. *Deconstructive Subjectivities*. New York: SUNY. Pp. 13–45.

Critchley, S. (1999) *Ethics, Politics, Subjectivity*. London: Verso.

Cumberland, R. (2005 [1672]) *A Treatise of the Laws of Nature*. Indianapolis: Liberty Fund.

Curran, J. (2002) 'Media and Making of British Society, c1700–2000,' *Media History*, 2(1): 135–154.

Damasio, A. (2011) *Self Comes to Mind: Constructing the Conscious Brain*. London: Vintage Books.

DeCew, J.W. (1997) *In Pursuit of Privacy: Law, Ethics, and the Rise of Technology*. New York: Cornell University Press.

Deleuze, G. (1988 [1970]) *Spinoza: Practical Philosophy*. San Francisco: City Lights.

Deleuze, G. (1992) 'Postscript on the Societies of Control,' *October*, 59(4): 3–7.

Deleuze, G. (2011 [1969]) *The Logic of Sense*. London: Continuum.

Deleuze, G. (2011 [1981]) *Francis Bacon: The Logic of Sensation*. London: Continuum.

Deleuze, G. and Guattari, F. (2011 [1994]) *What is Philosophy?* London: Verso.

Dennett, D.C. (1998 [1987]) *The Intentional Stance*. Cambridge, MA: MIT.

De Waal, F. (2012) *The Age of Empathy: Nature's Lessons for a Kinder Society*. London: Souvenir.

Dewey, J. (1995 [1908]) 'Does Reality Possess Practical Character,' in R.B. Goodman, ed. *Pragmatism: A Contemporary Reader*. New York: Routledge. Pp. 79–92.

Donald, M. (2001) *A Mind So Rare. The Evolution of Human Consciousness*. New York: W.W. Norton.

Dreyfus, H.L. (1991) *Being-in-the-World: A Commentary on Heidegger's Being and Time, Division 1*. Cambridge, MA: MIT.

Elmer, G. (2004) 'A Diagram of Panoptic Surveillance,' *New Media and Society*, 5(2): 231–247.

Elmer, G. (2012) 'Panopticon—Discipline—Control' in K. Ball, K. Haggerty and D. Lyon (eds.). *Routledge Handbook of Surveillance Studies*. Oxon: Routledge. Pp. 21–29.

Engelhardt, H.T. Jr. (2000) 'Privacy and Limited Democracy: The Moral Centrality of Persons' *Social Philosophy & Policy*, 17(2): 120–140.

Feuerbach, L. (1957 [1841]) *The Essence of Christianity*. New York: Harper & Row.

Ford, C.S. and Frank, F.A. (1951) *Patterns of Sexual Behavior*. New York: Harper and Brothers.

Foucault, M. (1977) *Discipline and Punish*. London: Penguin.

Foucault, M. (1990 [1976]) *The History of Sexuality Vol.1*. London: Penguin.

Frank, M. (1989) *What is Neostructuralism?* Minneapolis: University of Minnesota Press.

Fromm, E. (2002 [1956]) *The Sane Society*. Oxon: Routledge.

Fuchs, C. (2012) Dallas Smythe Today—The Audience Commodity, the Digital Labour Debate, Marxist Political Economy and Critical Theory. Prolegomena to a Digital Labour Theory of Value, *tripleC*, 10(2): 692–740.

Fuller, M. and Goffey, A. (2012) *Evil Media*. Cambridge, MA: MIT Press.

Gavison, R. (1984 [1980]) 'Privacy and the Limits of the Law,' in F.D. Schoeman, ed. *Philosophical Dimensions of Privacy: An Anthology*. Cambridge: Cambridge University Press. Pp. 346–402.

Giddens, A. (1984) *The Constitution of Society*. Polity Press: Cambridge.

Glenn, C. (1994) 'Sex, Lies, and Manuscript: Refiguring Aspasia in the History of Rhetoric,' *Composition and Communication*, 45(4): 180–199.

Goffman, E. (1990 [1959]) *The Presentation of the Self in Everyday Life*. London: Penguin.

Goldman, A.I. (2006) *Simulating Minds: The Philosophy, Psychology, and Neuroscience of Mindreading*. Oxford: Oxford University Press.

Grotius, H. (2001 [1625]) *On the Law of War and Peace*. Ontario: Batoche Books.

Habermas, J. (1990 [1985]) *The Philosophical Discourse of Modernity*. Cambridge, MA: MIT Press.

Halewood, M. and Michel, M. (2008) 'Being a Sociologist and Becoming a Whiteheadian: Toward a Concrescent Methodology,' *Theory, Culture & Society*, 25(4): 31–56.

Hall, E.T. (1969) *The Hidden Dimension*. New York: Anchor.

Hardt, M. and Negri, A. (2000) *Empire*. Cambridge, MA: Harvard University Press.

Harman, G. (2005) *Guerilla Metaphysics: Phenomenology and the Carpentry of Things*. Chicago: Open Court.

Harman, G. (2007) *Heidegger Explained: From Phenomenon to Thing*. Chicago: Open Court.

Harman, G. (2008) 'Delanda's Ontology: Assemblage and Realism,' *Continental Philosophy Review*, 41(3): 367–383.

Harman, G. (2009) *Prince of Networks: Bruno Latour and Metaphysics*. Melbourne: re.press.

Harman, G. (2010) *Towards Speculative Realism: Essays and Lectures*. Ropley: Zero Books.

Hartnack, J. (1952) 'The Alleged Privacy of Experience,' *Journal of Philosophy*, 49(12): 405–411.

Hayles, K. (1999) *How We Became Posthuman: Virtual Bodies in Cybernetics, Literature, and Informatics*. Chicago: University of Chicago Press.

Hayles, K. (2005) *My Mother Was a Computer: Digital Subjects and Literary Texts*. Chicago: University of Chicago Press.

Hegel, G.W.F. (1910 [1807]) *The Phenomenology of Mind*. New York: MacMillan.

Hegel, G.W.F. (2001 [1820]) *The Philosophy of Right*. Ontario: Batoche Books.

Heidegger, M. (1988 [1927]) *The Basic Problems of Phenomenology*. Bloomington, IN: Indiana University Press.

Heidegger, M. (1993 [1954]) 'The Question Concerning Technology,' in M. Heidegger *Basic Writings*, ed. D.F. Krell. New York: Harper Collins. Pp. 311–341.

Heidegger, M. (2011 [1962]) *Being and Time*. New York: Harper & Row.

Heidegger, M. (2012 [1936–8]) *Contributions to Philosophy (of the Event)*. Bloomington, IN: Indiana University Press.

Heidegger, M. (2013 [1941–2]) *The Event*. Bloomington, IN: Indiana University Press.

Herder, J.G. (2002 [1774]) 'This Too a Philosophy of History for the Formation of Humanity' in *Philosophical Writings*, ed. M.N. Forster. Cambridge: Cambridge University Press. Pp. 272–358.

Hier, S.P. (2003) 'Probing the Surveillant Assemblage: On the Dialectics of Surveillance Practices as Processes of Social Control,' *Surveillance & Society* 1(3): 399–411.

Hildebrandt, M. and Gutwirth, S, eds. (2008) *Profiling the European Citizen: Cross-Disciplinary Perspectives*. Dordrecht: Springer,

Hobbes, T. (1985 [1651]) *Leviathan*. London: Penguin.

Hobbes, T. (1998 [1668/1642]) *Man and Citizen (De Homine and De Cive)*. Bloomington, In: Hackett.

Home Office (2011) Crime in England and Wales 2010/11, https://www.gov.uk/government/uploads/system/uploads/attachment_data/file/116417/hosb1011.pdf, date accessed 30/06/13.

Hoque, M.E. (2012) *Exploring Temporal Patterns of Spontaneous Smiles When Delighted and Frustrated*, http://web.media.mit.edu/~mehoque/Temporal_Patterns_of_Smile.htm, date accessed 05/06/13.

Hume, D. (1965 [1739]) *A Treatise of Human Nature*. London: Oxford University Press.

Hume, D. (2005 [1753]) *Essays and Treatises on Several Subjects*. http://books.google.co.uk/books?id=vRPiea4s53QC, accessed 18/08/13.

Husserl, E. (1970 [1900]) *Logical Investigations*. London: Routledge.

Husserl, E. (1970 [1936]) *The Crisis of European Sciences and Transcendental Philosophy*. Evanston: Northwestern University Press.

Husserl, E. (2002 [1952]) *Ideas Pertaining to a Pure Phenomenology and to a Phenomenological Philosophy: Second Book*. Dordrecht: Kluwer.

Hutcheson, F. (2013 [1725]) Inquiry Concerning Moral Good and Evil, *Online Library of Liberty*, http://oll.libertyfund.org/?option=com_staticxt&staticfile=show.php%3Ftitle=2075&chapter=198065&layout=html&Itemid=27, date accessed 13/08/13.

James, W. (2000 [1907]) *Pragmatism and Other Writings*, ed. G. Gunn. New York: Penguin.

Kant, I. (1952 [1790]) *The Critique of Judgement*. Oxford: Oxford University Press.

Kant, I. (1983 [1784]) 'An Answer to the Question: What is Enlightenment?' in *Perpetual Peace and Other Essays*. Indianapolis: Hackett. Pp. 33–48.

Kant, I. (1983 [1793]) 'On the Proverb: That May be True I Theory, But Is of No Practical Use' in *Perpetual Peace and Other Essays*. Indianapolis: Hackett. Pp. 61–92.

Kant, I. (1990 [1781]) *The Critique of Pure Reason*. London: MacMillan.

Kant, I. (1991 [1785]) *The Moral Law: Grounding for the Metaphysics of Morals*. London: Routledge.

Kegan, S. (1992) 'The Structure of Normative Ethics,' *Philosophical Perspectives, Ethics*, 6: 223–242.

Kosinki, M., Stillwell, S. and Graepel, T. (2012) 'Private traits and attributes are predictable from digital records of human behavior' *Proceedings of the National Academy of Sciences of the United States of America*, http://www.pnas.org/content/early/2013/03/06/1218772110, date accessed 30/03/13.

Lanier, J. (2010) *You Are Not a Gadget*. London: Penguin.

Latour, B. (1987) *Science in Action*. Cambridge, MA: Harvard University Press.

Latour, B. (1988) *The Pasteurization of France*. Cambridge, MA: Harvard University Press.

Latour, B. (1992) 'The Sociology of a Few Mundane Artifacts,' in W.E. Bijker and J. Law, eds. *Shaping Technology Building Society*. Cambridge, MA: MIT Press. Pp. 225–258.

Latour, B. (1993) *We Have Never Been Modern*. Cambridge, MA: Harvard University Press.

Latour, B. (1996) *Aramis, or the Love of Technology*. Cambridge, MA: Harvard University Press.

Latour, B. (1999) *Pandora's Hope: Essays on the Reality of Science Studies*. Cambridge, MA: Harvard University Press.

Latour, B. (2004) *Politics of Nature: How to Bring the Sciences Into Democracy*. Cambridge, MA: Harvard University Press.

Latour, B. (2005) *Reassembling the Social: An Introduction to Actor-Network-Theory*. Oxford: Oxford University Press.

Latour, B. (2013) *An Inquiry Into Modes of Existence: An Anthropology of the Moderns*. Cambridge, MA: Harvard University Press.

Law, J. (2012 [1987]) 'Technology and Heterogeneous Engineering: The Case of Portuguese Expansion,' in W.E. Bijker, T.P. Hughes and T. Pinch, eds. *The Social Construction of Technological Systems: New Directions in the Sociology and History of Technology*. Cambridge, MA: MIT. Pp. 105–127w.

Lazzarato, M. (2004) 'From Capital-Labour to Capital-Life,' *Ephemera*, 4(3): 187–208.

Leveson, The Right Honourable Lord Justice (2012) *An Inquiry Into the Culture, Practices and Ethics of the Press: Executive Summary*. London: The Stationary Office.

Levinas, E. (1999 [1974]) *Otherwise Than Being, or, Beyond Essence*. Pittsburgh, PA: Duquesne University Press.

Lieshout, M.v., Kool, L., Schoonhoven, B.v. and Jonge, M. de. (2011) *Privacy by Design: Alternative for Existing Practices in Safeguarding Privacy?* Paper prepared for EuroCPR 2011. 1–29.

Lingis, A. (1998) *The Imperative*. Bloomington, IN: Indiana University Press.

Litt, E. (2102) 'Knock, Knock. Who's There? The Imagined Audience,' *Journal of Broadcasting & Electronic Media*, 56(3): 330–345.

Lloyd, J. (2012) 'Exposed: The 'Swaggering Arrogance' of the Popular Press,' in R.L. Keeble and J. Mair, eds. *The Phone Hacking Scandal: Journalism on Trial*. Bury St. Edmunds: Abramis. Pp. 1–3.

Locke, J. (2005 [1690]) *Second Treatise on Government and A Letter Concerning Toleration*. Stilwel, KS: Digireads.com

Lukács, G. (1967 [1923]) *History & Class Consciousness*. London: Merlin Press.

Lyon, D. (2001) *Surveillance Society: Monitoring Everyday Life*. Buckingham: Open University Press.

Lyotard, J.-F. (2004 [1974]) *Libidinal Economy*. London: Continuum.

Mackenzie, A. (2002) *Transductions: Bodies and Machines at Speed*. New York: Continuum.

MacKinnon, C. (1989) *Toward a Feminist Theory of the State*. Cambridge, MA: Harvard University Press.

Malinowski, B. and Ellis, H. (2005 [1929]) *The Sexual Life of Savages in North-Western Melanesia: An Ethnographic Account of Courtship, Marriage, and Family Life Among the Natives of the Trobriand Islands, British New Guinea*. Whitefish, MT: Kessinger.

Mann, S. (2013) *Veillance and Reciprocal Transparency: Surveillance Versus Sousveillance, AR Glass, Lifelogging, and Wearable Computing*, http://wearcam.org/veillance/veillance.pdf, date accessed 20/09/13

Mann, S., Nolan, J. and Wellman, B. (2003) 'Sousveillance: Inventing and Using Wearable Computing Devices for Data Collection in Surveillance Environments,' *Surveillance & Society*, 1(3), 331–355.

Marcuse, H. (1955 [1941]) *Reason and Revolution: Hegel and the Rise of Social Theory*. London: Routledge.

Marx, K. (1970 [1859]) *A Contribution to the Critique of Political Economy*. Moscow: Progress.

Marx, K. (1999 [1867]) *Capital: An Abridged Edition*. Oxford: Oxford University Press.

Marx, K. (2012 [1844]) 'On the Jewish Question' in K. Marx and F. Engels. Collected Works Vol. 3, http://www.marxists.org/archive/marx/works/cw/volume03/index.htm, date accessed 10/09/13.

Marx, K. and Engels, F. (2011 [1844]) *Economic and Philosophic Manuscripts of 1844*. Blacksburg, VA: Wilder.
McStay, A. (2011a) Profiling Phorm: An Autopoietic Approach to the Audience-as-Commodity. *Surveillance and Society*, 8(3), 310–322.
McStay, A. (2011b) *The Mood of Information: A Critique of Behavioural Advertising*. New York: Continuum.
McStay, A. (2013a) *Creativity and Advertising: Affect, Events and Process*. London: Routledge.
McStay, A. (2013b) I Consent: An Analysis of the Cookie Directive and its Implications for UK Behavioural Advertising, *New Media and Society*, 9(2), 187–211.
McStay, A. (2013c) *Open Rights Group Conference (ORG 2013)*, http://advertising-communications-culture.blogspot.co.uk/2013/06/open-rights-group-conference-org-2013.html, date accessed 05/09/13.
Mead, M. (2001 [1928]) *Coming of Age in Samoa*. New York: Harper Perennial.
Merleau-Ponty, M. (2002 [1945]) *Phenomenology of Perception*. London: Routledge.
Mill, J.S. (1962 [1838]) 'Bentham', in M. Warnock eds. *Utilitarianism, On Liberty, Essay on Bentham*. London: Fontana Press. Pp. 78–125.
Mill, J.S. (1962 [1859]) *Utilitarianism, On Liberty, Essay on Bentham*. London: Fontana Press.
Moore, B. Jr. (1984) *Privacy: Studies in Social and Cultural History*. New York: M.E. Sharpe.
Nissenbaum, H. (2004) 'Privacy as Contextual Integrity,' *Washington Law Review*, 119–158.
Nissenbaum, H. (2010) *Privacy in Context: Technology, Policy, and the Integrity of Social Life*. Stanford: Stanford University Press.
Petronio, S. (1991) 'Communication Boundary Management: A Theoretical Model of Managing Disclosure of Private Information Between Married Couples,' *Communication Theory*, 1: 311–335.
Petronio, S. (2002) *Boundaries of Privacy: Dialectics of Disclosure*. Albany, NY: SUNY Press.
Plato (1993 [360 BC]) *Sophist*. Indianapolis: Hackett.
Plato (2004 [360 BC]) *The Laws*. London: Penguin.
Posner, R.A. (1983) *The Economics of Justice*. Cambridge, MA: Harvard University Press.
Poster, M. (1990) *The Mode of Information: Poststructuralism and Social Context*. Chicago: University Of Chicago Press.
Poster, M. (1995) *The Second Media Age*. Cambridge, MA: Polity Press.
Privacy International (2013) *Who, What, Why*, available from https://www.privacyinternational.org/, date accessed 04/06/13.
Rawls, J. (2005 [1993]) *Political Liberalism*. New York: Columbia University Press.
Rorty, R. (1989) *Contingency, Irony and Solidarity*. Cambridge: Cambridge University Press.
Rorty, R. (1998) *Truth and Progress: Philosophical Papers 3*. Cambridge: Cambridge University Press.
Rorty, R. (2007) *Philosophy as Cultural Politics: Philosophical Papers 4*. Cambridge: Cambridge University Press.
Rorty, R. (2010 [1993]) 'Human Rights, Rationality, and Sentimentality' in C.J. Voparil and R.J. Bernstein, eds. *The Rorty Reader*. Chichester: Wiley-Blackwell. Pp. 351–365.
Rosenberg, A. (2000) 'Privacy as a Matter of Taste and Right,' *Social Philosophy & Policy*, 17(2): 68–90.
Rössler, B. (2005) *The Value of Privacy*. Cambridge: Polity Press.

Rousseau, J.J. (2004 [1755]) *Discourse on the Origin and Foundations of Inequality Among Men*, http://www.gutenberg.org/files/11136/11136.txt, date accessed 04/06/13.

Rousseau, J.J. (2008 [1762]) *The Social Contract*. New York: Cosimo.

Rusbridger, A. (2012) 'Hackgate "Reveals failure of Normal Checks and Balances to Hold Power to Account",' in R.L. Keeble and J. Mair, eds. *The Phone Hacking Scandal: Journalism on Trial*. Bury St. Edmunds: Abramis. Pp. 129–144.

Russell, B. (1958) 'What is Mind?,' *The Journal of Philosophy*, 55(1): 5–12.

Ryle, G. (2000 [1949]) *The Concept of Mind*. Penguin: London.

Sartre, J.-P. (2003 [1943]) *Being and Nothingness*. Oxon: Routledge.

Schmidt, E. and Cohen, J. (2013) *The New Digital Age: Reshaping the Future of People, Nations and Business*. London: John Murray.

Schneewind, J.B. (1998) *The Invention of Autonomy: A Hitory of Moral Philosophy*. Cambridge: Cambridge University Press.

Schoeman, F.D. (2008 [1992]) *Privacy and Social Freedom*. Cambridge: Cambridge University Press.

Searle, J.R. (1983) *Intentionality: An Essay in the Philosophy of Mind*. Cambridge: Cambridge University Press.

Searle, J.R. (1996) *The Construction of Social Reality*. London: Penguin.

Searle, J.R. (1998) *The Mystery of Consciousness*. London: Granta Publications.

Simmel, G. (1906) 'The Sociology of Secrecy and of Secret Societies,' *American Journal of Sociology*, 11: 441–498.

Skinner, B.F. (1976) *About Behaviourism*. New York: Vintage.

Smith, A. (2011 [1759]) *Theory of Moral Sentiments*. Kapaau: Gutenberg Publishers.

Smythe, D.W. (1977) 'Communications: Blindspot of Western Marxism,' *Canadian Journal of Political and Social Theory*, 1 (3), 1–27.

Smythe, D.W. (1981) 'On the Audience Commodity and its Work' in *Dependency Road: Communications, Capitalism, Consciousness, and Canada* (Norwood, NJ: Ablex).

Spinoza, B. (1996 [1677]) *Ethics*. London: Penguin.

Stiegler, B. (2010) *For a New Critique of Political Economy*. Cambridge: Polity.

Tarde, G. and Durkheim, E. (2010 [1903]) 'The Debate in M. Candea.' *The Social After Gabriel Tarde: Debates and Assessments*. Abingdon, Oxon: Routledge. Pp. 27–43.

Terranova, T. (2007) 'Futurepublic: On Information Warfare, Bio-Racism and Hegemony as Noopolitics,' *Theory, Culture & Society*, 24(3): 125–145.

Thompson, E.P. (1963) *The Making of the English Working Class*. London: Gollancz.

Thrift, N. (2008) *Non-Representational Theory: Space, Politics, Affect*. London: Routledge.

Turkle, S. (2011) *Alone Together: Why We Expect More From Technology and Less From Each Other*. New York: Basic Books.

US Patent & Trademark Office (2012) *Methods and Systems for Presenting an Advertisement Associated With an Ambient Action of a User*, available from http://appft1.uspto.gov/netacgi/nph-Parser?Sect1=PTO2&Sect2=HITOFF&p=1&u=%2Fnetahtml%2FPTO%2Fsearch-bool.html&r=1&f=G&l=50&co1=AND&d=PG01&s1=20120304206&OS=20120304206&RS=20120304206, date accessed 07/02/13.

Vattimo, G. (1988) *The End of Modernity*. Cambridge: Polity Press.

von Neumann, J. (1963) *John von Neumann: Collected Works*. New York: Pergamon Press.

Waldron, J. (1987) 'The Theoretical of Liberalism,' *The Philosophical Quarterly*, 37(147): 127–150.

Warren, S. and Brandeis, L. (1984 [1890]) 'The Right to Privacy [The Implicit Made Explicit],' in F.D. Schoeman, ed. *Philosophical Dimensions of Privacy: An Anthology*. Cambridge: Cambridge University Press. Pp. 75–103.

Westin, A. (1984 [1967]) 'The Origins of Modern Claims to Privacy, in F.D. Schoeman, ed. *Philosophical Dimensions of Privacy: An Anthology*. Cambridge: Cambridge University Press. Pp. 56–74.

Whitehead, A.N. (1948 [1933]) *Adventures of Ideas*. New York: Free Press.

Whitehead, A.N. (1964 [1920]) *The Concept of Nature*. Cambridge: Cambridge University Press.

Whitehead, A.N. (1968 [1938]) *Modes of Thought*. New York: Free Press.

Whitehead, A.N. (1985 [1929]) *Process and Reality: Corrected Edition*. New York: Free Press.

Whitehead, A.N. (1997 [1925]) *Science and the Modern World*. New York: Free Press.

Williams, R. (1992) *Television: Technology and Cultural Form*. Hanover, CT: Wesleyan University Press.

Winograd, T. and Flores, F. (1987) *Understanding Computers and Cognition: A New Foundation for Design*. Norwood, NJ: Ablex.

Wittgenstein, L. (2007 [1921]) *Tractatus Logico-Philosophicus*. Champaign, IL: Standard.

Wittgenstein, L. (2009 [1953]) *Philosophical Investigations*. Chichester: Wiley-Blackwell.

Wood, D. (2003) 'Editorial. Foucault and Panopticism Revisited,' *Surveillance & Society*, 1(3): 234–239.

Zimmerman, M.E. (1990) *Heidegger's Confrontation With Modernity: Technology, Politics, Art*. Bloomington, IN: Indiana University Press.

Index

A

actant 10, 88–101, 137, 140–142, 146–148, 154, 155
advertising 8, 9, 20, 59, 71, 75, 77–81, 86, 87, 105, 117, 131, 149, 155
affect 4–12, 73, 83, 84, 88, 93–101, 105–110, 124, 131, 134–150, 158–162
affective breach 138, 163
agency 28, 32, 34, 41, 108, 110, 120, 137
   (also see autonomy)
aggregation 149
a-historical data mining 64, 69, 75, 82, 163
alienation 11, 123–133, 141, 157, 158
Archimedean p$2rinciple 35, 41, 42, 56, 67
Arendt, Hannah 7, 16, 19–21
Aristotle 15, 17, 18, 21, 66, 68, 107, 151
artificial intelligence 76, 86
assemblages 10, 11, 73, 86, 91, 92, 94–101, 103, 108, 110, 118, 125, 131, 136, 143, 145–149

autonomy 1, 2, 7, 22–24, 29–35, 41, 42, 45, 47, 51, 52, 55–57, 64, 68, 70, 91, 110, 120, 123, 126, 139, 152, 157, 164
   oppressive 41, 164
autopoiesis 9, 78, 79

B

Bateson, Gregory 9, 77, 78, 117, 121
Bayesian learning 85, 149, 154
Beccaria, Cesare 37, 46
behavior 1, 4, 5, 10, 15, 20, 41, 42, 52–54, 99, 119, 137, 138–141, 157, 158, 161, 162, 165
   advertising (see advertising)
   educate 28
   unaccountable 2
   mining 9, 64, 69, 71, 74–86, 108–117, 154
behaviorism 75–77, 112, 115, 118, 119, 134, 156

being alone 2, 3, 4, 5, 22, 28, 35, 39, 42, 43, 45, 47, 139, 152, 165 (also see *seclusion trap, isolation and seclusion*)
Bentham, Jeremy 8, 9, 36–39, 41, 43, 44, 46, 47, 55, 56, 65, 111, 152
Berardi, Franco 124, 157
Berlin, Isaiah 27–29, 40, 41
Berkeley, George (Bishop) 115
big data 23
biopower (see power)
black boxes 10, 95–98, 145, 147, 148, 161
    privacy as 101, 140
Bobbio, Norberto 43
body-doubles 110, 163
borders 1, 20, 21, 52, 97, 113, 114, 138, 148, 149
Brentano, Franz 10, 71, 72, 102–104, 155

C

capitalism 20, 54, 127–131
    anti- 140
carnality 105, 106, 147
citizenship 16, 17, 30, 123, 125
Clarke, Samuel 24, 25
co-evolving authorship 9, 78, 79, 134, 163
Comte, Auguste 29
community 5, 16, 25, 35, 37, 42, 55, 83, 94, 123, 126, 132, 152, 153, 157–159, 164
    facts 159, 163 (also see privacy norms)
concealing 19, 45, 67, 68, 75, 165
consent 22, 23, 25–27, 31–36, 40, 50, 63, 152, 153, 159, 160
contextual 8, 49–52, 58, 59, 71, 94, 98, 101, 143, 145, 149, 153, 158, 159, 164
control (see privacy)
co-production 78, 145
Cumberland, Richard 36, 37, 46, 47, 152
Cybernetics 51, 89

D

Damasio, Antonio 120
Dennett, Daniel 76, 104
data mining 9, 40, 64, 68–70, 73, 77, 86, 109, 115, 124, 128, 131, 152, 153, 163, 164
dispositifs 56, 135
dispositional competence 74, 76, 163
Deleuze, Gilles 6, 135–137, 142, 146, 158
deontic 8, 31, 37, 49, 53, 55, 59, 64, 152 (also see deontology)
deontology 6, 29, 50, 160
Derrida, Jacques 56, 108
Descartes, René 107, 112
determinism 25, 28, 68, 95, 136, 153
Dewey, John 8, 49, 54, 56
data protection 23, 33, 59
dualism 3, 6, 7, 16, 21, 34, 47, 53, 91, 111, 112, 121–123, 136–138, 151
Durkheim, Émile 99

E

ecology 4, 5, 78, 93, 95, 99, 121, 149, 155, 156, 161, 164 (also see *political ecology of privacy*)
efficiency 8, 34, 36, 46, 55, 66
emergent protocol (see protocol)
Empathic machines/media 11, 74, 82, 84–87, 108, 114–117, 119, 122, 128, 131, 149, 155–157, 163, 164
Empathy 9, 76, 82–87, 102, 106, 149, 155
epistemology 7, 8, 134, 151 (also see knowledge)
Engels, Friedrich 128, 129, 157
Enlightenment 8, 24, 27, 29–31, 36, 44, 45, 55, 58, 165
estrangement (also see alienation) 124, 128
ethnocentricism 15, 50, 55, 57, 67, 90, 138, 153, 159–161, 163, 165
Europe 23, 33, 47, 53, 59, 160
Events 9, 71, 73, 74, 79, 83, 89, 96, 98, 101, 102, 106, 109, 113, 144, 151

INDEX | 183

affective 139, 144, 161
Heidegger 9, 63–65, 67, 68, 71–76, 79, 153
privacy 4, 32, 51, 137–139, 140, 143, 144, 161, 162, 164
Whitehead 7, 12, 97, 143–150, 158
everyday mutuality 35, 164

F

Facebook 4, 113, 130, 132, 140, 141
Feminism 2
Feuerbach, Ludwig 127
Fichte, Johann 29, 126
Foucault, Michel 11, 43, 56, 57, 59, 135, 142
foundationalism 6, 8, 55, 57, 96
freedom 1, 7, 16, 22–24, 26–34, 38–40, 51, 64, 91, 110, 123–126, 129, 139, 141, 152
Fromm, Erich 130

G

Gadamer, Hans-Georg 82
gay (see homosexuality)
Goffman, Erving 52
Google 4, 132, 138
Grotius, Hugo 25, 30

H

Habermas, Jurgen 20, 56, 59, 142
Hardt, Michael 124, 130, 157
Harman, Graham 9, 10, 65, 96, 103–106, 108, 155, 161
Hegel, Georg 6, 11, 56, 68, 69, 124–127, 131, 132, 133, 157
Heidegger, Martin 6, 8–10, 47, 56, 63–77, 79, 81, 83, 86, 88–91, 93, 101, 103, 108, 128, 130, 131, 144, 153–155, 161, 164
Helvetius, Claude 37, 46
Herder, Johann 83

Heterogeneous 76, 78, 79, 86, 90, 93, 109, 154, 163, 164
Hobbes, Thomas 25–27, 37, 124
homosexuality 18, 42
Hume, David 25, 33, 82, 83
Husserl, Edmund 10, 71, 72, 77, 78, 82, 83, 102, 104, 105, 155
Hutcheson, Francis 46
hybrids (Latourian) 10, 88, 90–96, 98, 155

I

immanence 93, 98, 121, 136–138, 142, 158
intentionality 10, 75, 77, 102–106, 155
  machinic 10, 164
intimacy 1, 77
isolation 42, 42, 152, 159 (also see being alone and *seclusion trap*)

J

James, William 8, 49, 51, 54
Journalism 3

K

Kant, Immanuel 6, 23, 24, 29–32, 34, 40, 47, 50, 55, 72, 78, 91, 92, 96, 103–107, 126, 127, 132, 152, 161
knowledge (possibilities of machinic knowing) 7, 8, 11, 63, 72, 74, 85, 106, 115, 117, 119, 134 (also see epistemology)

L

labour (immaterial) 157
Latour, Bruno 6, 10, 11, 88–101, 104, 115, 136, 142, 143, 145–148, 154, 155, 161
Lazzarato, Maurizio 124, 130, 157

Leveson 148
liberalism 2, 3, 6, 7, 11, 21–25, 29, 31–36, 39, 41, 43, 44, 46, 49, 51, 55–58, 63, 108, 110, 113, 120, 140, 152, 155–157, 159, 164, 165
   anti- 64, 68, 70, 123, 124, 126, 132
lifesharing 21
Lingis, Alphonso 105, 106
local realism 94, 164
Locke, John 22, 24, 27, 32, 124
Lukács, György 130
Lyotard, Jean-François 131

M

machinic verisimilitude 9, 75–77, 81, 83, 85, 87, 154, 163–165
machinism 66, 82
Marcuse, Herbert 124–126, 131
Marxism 2, 11, 44, 68, 72, 97, 99, 123, 124, 126, 128–133, 157, 158–160
Marx, Karl 56, 69, 123, 124, 127–130, 132, 133
Merleau-Ponty, Maurice 103–105
metadata 148
metaphysics 10, 76, 89, 96, 98, 107, 119, 139, 142, 154, 160, 161
   morality 37, 49, 56–58
      presence (Heideggerian) 8, 47, 63, 66–70, 73, 79, 88, 153
Mill, John Stuart 8, 22–24, 28, 30, 34, 37–42, 47, 49, 56, 152
Moods 72, 74, 77, 83, 109, 116, 118, 131, 154, 164
mood of information 9, 44, 75–79, 81, 82, 109, 117, 131, 164
morality 24, 25, 29, 31, 32, 34, 35, 37, 39, 45, 49, 55, 58, 126

N

neutral monism 136
Negri, Antonio 124, 130, 157

Nietzsche, Friedrich 56, 70, 108
Nissenbaum, Helen 8, 50, 51, 53, 59, 143, 153
noopolitics 21
normative 8, 29, 50, 51, 53, 54, 160, 165

O

oikos 16–18
oppressive autonomy (see autonomy)

P

paternalism 18, 28, 29, 35, 40, 53, 59
phenomenal materialism 135, 164
Plato 18, 19, 21, 56, 66, 68, 152
policy 6, 10, 23, 32, 55, 56, 59, 98, 101, 110, 153, 155, 159, 160
polis 16–18
political ecology of privacy 95, 164
Posner, Richard 8, 36, 39, 44–47, 65, 157
Poster, Mark 110, 164
post-human 111
power 2, 21, 22, 26, 31, 39, 40, 43, 44, 47, 48, 100, 128–131, 135, 165
   bio 11, 56, 57
   criticisms 100
   will to 68
pragmatism 6–8, 49–51, 54, 55, 92, 94, 152, 153, 159
prehensions (Whiteheadian) 105, 145–147
privacy
   by design 90
   control 1, 2, 7, 16, 23, 40, 45, 52, 56, 122, 124, 131, 135, 149, 152, 153
   (also see liberalism and privacy management)
   informational 8, 35, 51, 53, 110, 117, 124, 149, 154, 159
   management 4, 6, 30, 43, 45, 52, 113, 143, 152

norms 7, 8, 15, 34, 49–55, 59, 90, 94, 95, 99, 101, 115, 122, 138, 140, 143, 149, 153, 158–161, 163 (also community facts)
  protocol 4, 5, 8, 10, 11, 34, 35, 52–54, 63, 93–95, 98, 99, 101, 113, 134, 135, 137–142, 144, 145, 150, 151–153, 158–162, 163–165
  scripts 5, 98, 147, 148, 165
Privacy International 1, 2
protocol (see privacy)
public traces 21, 109, 110, 116, 117, 122, 124, 128, 131, 132, 157, 160, 163, 165

Q

qualified right 2, 97

R

radical transparency (see transparency)
rationalism 24, 41, 51, 58, 69, 70, 75, 153
Rawls, John 43
reason (philosophy) 24, 25, 28, 29, 31, 37, 41, 42, 45, 47, 49, 50, 58, 86, 107, 118, 126, 127, 152
redescription 50, 57, 58
relationships (with others) 1, 2, 5, 8, 9, 10, 17, 18, 35, 40, 43, 52, 78, 120, 125, 144, 148, 149 (also see privacy management)
relativism 45, 58, 158, 159
rights 1, 2, 3, 8–11, 18, 22, 23, 25–45, 49, 53, 55, 58, 95–97, 101, 106, 122–126, 152, 155, 157, 159, 165
  to be left alone (see being alone and *seclusion trap*)
Rorty, Richard 8, 49, 53–59, 94, 100, 143
Rousseau, Jean-Jacques 27, 47
Russell, Bertrand 11, 119, 122
Ryle, Gilbert 6, 11, 75, 76, 111–121, 156

S

Sartre, Jean-Paul 140, 141, 158
scripts (see *privacy scripts*)
Searle, John 29, 76, 105, 168
seclusion 5, 17, 22, 28, 34, 35, 40, 42, 43, 47, 52, 64, 113, 125, 152, 164 (also see being alone)
  trap 40, 47, 165
self-determinism 24, 27, 33, 41, 55, 59, 70
sentiment mining 79, 81, 85, 113, 116, 117, 122, 124, 131
sex (acts) 1, 18
sexuality 18, 42, 46, 56, 113
Skinner, Burrhus 112, 115, 118, 156
Smith, Adam 82, 84
Smythe, Dallas 11, 78
Snowden, Edward 21, 22, 48, 52, 98, 145
social constructionism 32, 91, 95
sousveillance 21, 171
Spinoza, Baruch 6, 11, 25, 54, 134–138, 142, 145, 146, 158
Stiegler, Bernard 131,
substance-based fallacy 143, 165
substantialism 6, 10, 95, 145
surveillance 2, 4, 6, 20–23, 41, 44, 48, 52, 56, 69, 87, 94, 106, 110, 124, 135, 137, 139, 140, 145, 152, 155, 159, 164, 165
surveiller affect 142, 158, 165

T

taboo 1
technically negotiated privacy 165
teleology 8, 46
third-person being 140, 165
transduction 10, 94, 149, 158
transparency 2, 8, 9, 36, 43–48, 65, 69, 70, 140, 152, 165
  radical 8, 36, 44, 46, 47, 64, 65, 140, 152, 160, 165
Tarde, Gabriel 99, 101
Turing test 76

U

unconceal (see conceal)
utilitarian 6–8, 28, 36–51, 53, 55, 58, 152

V

verisimilitude 9, 75–77, 81, 83, 85, 87, 154, 163–165
Verizon 79–81, 87
Virno, Paolo 157
visceral 134, 135, 137, 161–163 (also see affect)
voyeurism 139, 141, 158, 165

W

Whitehead, Alfred 9, 12, 95, 97, 98, 104, 105, 143–150, 154, 155, 158
Wittgenstein, Ludwig 6, 11, 75, 76, 119–121, 156

Z

zombie media 82, 165

General Editor: **Steve Jones**

Digital Formations is the best source for critical, well-written books about digital technologies and modern life. Books in the series break new ground by emphasizing multiple methodological and theoretical approaches to deeply probe the formation and reformation of lived experience as it is refracted through digital interaction. Each volume in **Digital Formations** pushes forward our understanding of the intersections, and corresponding implications, between digital technologies and everyday life. The series examines broad issues in realms such as digital culture, electronic commerce, law, politics and governance, gender, the Internet, race, art, health and medicine, and education. The series emphasizes critical studies in the context of emergent and existing digital technologies.

Other recent titles include:

Felicia Wu Song
 Virtual Communities: Bowling Alone, Online Together

Edited by Sharon Kleinman
 The Culture of Efficiency: Technology in Everyday Life

Edward Lee Lamoureux, Steven L. Baron, & Claire Stewart
 Intellectual Property Law and Interactive Media: Free for a Fee

Edited by Adrienne Russell & Nabil Echchaibi
 International Blogging: Identity, Politics and Networked Publics

Edited by Don Heider
 Living Virtually: Researching New Worlds

Edited by Judith Burnett, Peter Senker & Kathy Walker
 The Myths of Technology: Innovation and Inequality

Edited by Knut Lundby
 Digital Storytelling, Mediatized Stories: Self-representations in New Media

Theresa M. Senft
 Camgirls: Celebrity and Community in the Age of Social Networks

Edited by Chris Paterson & David Domingo
 Making Online News: The Ethnography of New Media Production

To order other books in this series please contact our Customer Service Department:

(800) 770-LANG (within the US)
(212) 647-7706 (outside the US)
(212) 647-7707 FAX

To find out more about the series or browse a full list of titles, please visit our website:
WWW.PETERLANG.COM